HOLT

Biology

Study Guide

HOLT, RINEHART AND WINSTON

A Harcourt Education Company

Orlando • **Austin** • New York • San Diego • London

ISBN-13: 978-0-03-093223-6
ISBN-10: 0-03-093223-8

3 4 5 6 018 11 10 09 08

Contents

Contents

Contents

Contents

Skills Worksheet

Vocabulary Review

In the space provided, explain how the terms in each pair differ in meaning.

1. observation, hypothesis

_____

_____

_____

2. control group, experimental group

_____

_____

_____

3. hypothesis, theory

_____

_____

_____

4. homeostasis, metabolism

_____

_____

_____

5. heredity, reproduction

_____

_____

_____

6. heredity, evolution

_____

_____

_____

| Vocabulary Review *continued*

In the space provided, write the letter of the correct definition of the term.

_____ 7. biology

_____ 8. evolution

_____ 9. homeostasis

_____10. cell

_____11. reproduction

_____12. experiment

_____13. SI

_____14. metabolism

_____15. skepticism

a. the International System of Units, which is the measurement system that is accepted by scientists throughout the world

b. the maintenance of a constant internal state in a changing external environment

c. the study of living things

d. a habit of mind in which a person questions the validity of accepted ideas

e. the process of making offspring

f. the smallest unit that can perform all life functions

g. the process by which inherited characteristics of a species of organisms change over generations

h. a procedure carried out under controlled conditions to discover, demonstrate, or test an idea or explanation for something in the natural world

i. the sum of all of the chemical reactions that occur in an organism

Skills Worksheet

Test Prep Pretest

In the space provided, write the letter of the description that best matches the term or phrase.

_____ 1. hypothesis

_____ 2. experiment

_____ 3. observation

_____ 4. skepticism

_____ 5. theory

a. a system of ideas that explains many related observations and is supported by a large body of evidence

b. the act of perceiving objects or events by using the senses

c. an explanation that can be tested with observation, experimentation, or both

d. a procedure that is carried out under controlled conditions to test something in a scientific way

e. an attitude that involves questioning and doubt

Complete each statement by writing the correct term or phrase in the space provided.

6. The official name for the _____ _____

   is the International System of Units, which has the abbreviation _____.

7. _____ is the study of life.

8. The study of microscopic organisms is called _____. Scientists

   working in this field use _____

   _____ to grow microorganisms without contamination.

9. _____ is the study of animals. Scientists working in this field

   might collect data _____ using attached electronic devices.

10. The smallest unit capable of life is the _____. To be able

    to see most of these structures, you need a(n) _____.

11. The sum of all chemical processes an organism carries out is called the

    organism's _____.

12. Living organisms are able to maintain a constant state internally even though

    the external environment is always changing. This is called

    _____.

In the space provided, write the letter of the term or phrase that best completes each statement or best answers each question.

_____13. Which of these is *not* a key step in practicing scientific thought?
    a. observing small details in the world of nature
    b. supporting all conclusions with evidence
    c. accepting the opinions of any experienced scientist
    d. being willing to change ideas based on new discoveries

_____14. The first step in a scientific investigation is
    a. forming a hypothesis.   c. constructing a theory.
    b. running an experiment.   d. making observations.

_____15. Control groups and experimental groups are identical except for the
    a. group size.   c. dependent variable.
    b. independent variable.   d. conclusions.

_____16. In the International System of Units, measurements are scaled in multiples of
    a. 2.   c. 10.
    b. 5.   d. 1,000.

_____17. If you wanted to measure how much water a maple seedling needs each month to survive, which of these units would you be most likely to use?
    a. meters   c. milligrams
    b. liters   d. kilometers

_____18. Genetics is most closely related to the study of
    a. heredity.   c. reproduction.
    b. homeostasis.   d. responsiveness.

_____19. What properties do all living things exhibit?
    a. cellular organization, metabolism, homeostasis, reproduction, and heredity
    b. multicellular organization, metabolism, homeostasis, reproduction, and heredity
    c. metabolism, homeostasis, uniqueness, reproduction, and heredity
    d. cellular organization, homeostasis, reproduction, heredity, and variability

_____20. The process by which inherited characteristics present in a species change over generations is called
    a. growth.   c. predation.
    b. development.   d. evolution.

Read each question, and write your answer in the space provided.

21. What are three reasons it is important that scientific investigations be conducted ethically?

_____

_____

_____

22. Summarize three ways in which understanding science can help any person.

_____

_____

_____

23. Contrast the two different types of scientific experiments.

_____

_____

_____

24. How does the scientific use of the word *theory* compare with the common use of this word?

_____

_____

_____

25. Of the many rules for staying safe in a science lab, list the three that you think are most important. Give the reasons you chose each safety rule you listed.

_____

_____

_____

_____

26. How are reproduction and heredity similar? How are they different?

_____

_____

_____

Name _____ Class _____ Date _____

Vocabulary Review

In the space provided, explain how the terms in each pair differ in meaning.

1. genetics, genetic engineering

_____

_____

_____

2. epidemiology, vaccination

_____

_____

_____

3. ecology, environmental science

_____

_____

_____

In the space provided, write the letter of the description that best matches each term.

_____ 4. biometrics

_____ 5. ecology

_____ 6. environmental science

_____ 7. epidemiology

_____ 8. genetic engineering

_____ 9. genetics

_____ 10. genome

a. uses techniques such as geographic information systems (GIS) to map data

b. process by which a gene from a soil bacterium was transferred to corn

c. describes the complete set of hereditary information for an organism

d. the study of how organisms interact with each other

e. the study of how traits are passed from parents to offspring

f. uses methods such as DNA fingerprinting to determine a person's identity

g. involves studies such as Colwell's work with copepods and cholera pathogens

Skills Worksheet

Test Prep Pretest

In the space provided, write the letter of the term that best completes each statement or best answers each question.

_____ 1. Rita Colwell's work is an example of
- a. biotechnology.
- b. ecology.
- c. epidemiology.
- d. genetics.

_____ 2. A new bandage that helps prevent heavy blood loss was developed by
- a. battlefield medicine.
- b. satellite technology.
- c. nanotechnology.
- d. biotechnology.

_____ 3. All of the following are examples of biometrics *except*
- a. iris scanning.
- b. DNA fingerprinting.
- c. fingerprinting.
- d. satellite tagging.

_____ 4. New products that are based on biological structures or processes are developed through
- a. biometrics.
- b. biomimetics.
- c. genetics.
- d. nanotechnology.

_____ 5. What is the name for the study of the interrelationships among living organisms?
- a. biomimetics
- b. bioterrorism
- c. ecology
- d. epidemiology

In the space provided, write the letter of the description that best matches each term.

_____ 6. genetics

_____ 7. genome

_____ 8. biometrics

_____ 9. biomolecules

_____ 10. cloning

a. uses biological traits to identify individuals

b. the name for the complete set of hereditary information for an organism

c. may be used to produce extinct animals from their DNA

d. the science of heredity and the study of how traits are passed to offspring

e. organic compounds put together by organisms

Complete each statement by writing the correct term or phrase in the space provided.

11. An agent that causes a disease is called a(n) _____.

12. The disease _____ was conquered by vaccinations.

13. Bionic limbs and devices that help people hear, see, or talk are examples of

_____ _____.

14. The _____ _____

_____ has succeeded in sequencing all human chromosomes.

15. Biologists referred to the area of New Guinea discovered and explored in early

2006 as a(n) _____ _____.

Read each question, and write your answer in the space provided.

16. Explain how Rita Colwell used satellite data to predict cholera outbreaks.

_____

_____

_____

17. What are some ways in which biologists are working to eliminate diseases that affect human populations?

_____

_____

_____

18. Describe a "low-tech" solution Indian women use to prevent cholera.

_____

_____

19. How is biotechnology used in agriculture?

_____

_____

_____

20. Briefly describe how insulin for diabetics is currently produced.

_____

_____

_____

21. List some uses of computerized axial tomography (CAT) scanning.

_____

_____

_____

22. What are two examples of ethical concerns that are limiting research that involves biotechnology?

_____

_____

_____

_____

23. Name two kinds of living organisms that were among the new species discovered in New Guinea.

_____

_____

24. Why do biologists collect DNA samples from endangered wildlife?

_____

_____

_____

25. How can students make contributions to research in environmental science?

_____

_____

_____

_____

Name _____ Class _____ Date _____

Vocabulary Review

In the space provided, write the letter of the description that best matches each term.

_____ 1. ion

_____ 2. atom

_____ 3. compound

_____ 4. amino acids

_____ 5. buffer

_____ 6. pH

_____ 7. element

_____ 8. solution

_____ 9. molecule

a. smallest unit of matter that cannot be broken down by chemical means

b. a substance made of the joined atoms of two or more different elements

c. atom or molecule that has lost or gained one or more electrons

d. a substance made of only one type of atom

e. one substance evenly distributed in another

f. substance that prevents pH changes

g. building blocks of protein

h. measure of how acidic or basic a solution is

i. group of atoms held together by covalent bonds

Complete each statement by writing the correct term or phrase in the space provided.

10. A(n) _____ is a substance on which an enzyme acts during a

chemical reaction.

11. An organic compound with a ratio of one carbon atom to two hydrogen atoms to

one oxygen atom is a(n) _____.

12. Atoms tend to combine with each other in a way that each atom will have eight

_____ _____.

13. A(n) _____ is an organic compound that is not soluble

in water.

14. A(n) _____ is a long chain of amino acids.

15. Subunits of DNA and RNA are called _____.

16. DNA is a(n) _____ _____ that carries

genetic information.

| Vocabulary Review *continued*

In the space provided, explain how the terms in each pair differ in meaning.

17. acid, base

_____

_____

_____

18. cohesion, adhesion

_____

_____

_____

19. enzyme, active site

_____

_____

_____

_____

20. energy, activation energy

_____

_____

_____

21. DNA, RNA

_____

_____

_____

22. ATP, carbohydrate

_____

_____

_____

23. reactant, product

_____

_____

_____

Skills Worksheet

Test Prep Pretest

In the space provided, write the letter of the term or phrase that best completes each statement or best answers each question.

_____ 1. Acids and bases differ in that
   a. bases dissolved in water form more hydronium ions than do acids dissolved in water.
   b. acids dissolved in water form more hydronium ions than do bases dissolved in water.
   c. acids dissolved in water form more hydroxide ions than do bases dissolved in water.
   d. bases have a lower pH than do acids.

_____ 2. Which of the following groups of terms is associated with carbohydrates?
   a. monosaccharide, glycogen, cellulose
   b. monosaccharide, cellulose, lipid
   c. disaccharide, polysaccharide, steroid
   d. polysaccharide, amino acid, ATP

_____ 3. The speed of a chemical reaction is increased by
   a. an enzyme.   c. ATP.
   b. the reactant.   d. All of the above

_____ 4. Which of the following particles is found in an atom's nucleus?
   a. electron   c. isotope
   b. electron cloud   d. proton

_____ 5. Which of the following is *not* a property of water?
   a. cohesion   c. nonpolarity
   b. polarity   d. stores heat well

In the space provided, write the letter of the description that best matches each term.

_____ 6. atom
_____ 7. element
_____ 8. compound
_____ 9. molecule
_____ 10. electron

   a. a substance made when atoms of two or more different elements join together
   b. negatively charged atomic particle
   c. the smallest unit of matter that cannot be broken down by chemical means
   d. a group of atoms held together by covalent bonds
   e. a substance made of only one kind of atom

Complete each statement by writing the correct term or phrase in the space provided.

11. Molecules that are _____ dissolve best in water, while

_____ molecules do not dissolve well in water.

12. The weak chemical attractions between water molecules are

_____ bonds, while the stronger chemical bonds between the

atoms of each water molecule are _____ bonds.

13. An atom or a molecule that has gained or lost one or more electrons is called a(n)

_____.

14. On the pH scale, vinegar is a(n) _____ and ammonia is a(n)

_____.

15. Two _____ that store energy are starch, which is produced

by plants, and glycogen, which is produced by animals.

16. DNA and RNA, which are two kinds of _____

_____, are made of long chains of nucleotides.

17. When wood burns, a chemical reaction occurs in which the

_____ in wood are the reactants, and carbon dioxide and

water vapor are the _____.

18. Enzymes lower the _____ _____ of a

chemical reaction by holding the reactants close together in the right orientation.

19. A substrate attaches to the _____ _____

of an enzyme.

20. Temperature and _____ can affect enzyme activity.

Test Prep Pretest *continued*

Questions 21–23 refer to the figures below.

A.

B.

C.

21. Identify the class of organic compound represented by each of the molecules shown above.

_____

_____

22. For each type of compound shown above, explain the role it plays in your body.

_____

_____

_____

_____

_____

23. What are three characteristics of biomolecules?

_____

_____

_____

_____

_____

Test Prep Pretest *continued*

24. Explain why living things need energy and where they get it.

_____

_____

_____

_____

_____

25. Briefly describe the function of ATP in cells.

_____

_____

_____

_____

Skills Worksheet

Vocabulary Review

Complete each statement by writing the correct term or phrase from the list below in the space provided.

biodiversity	community	phosphorus cycle
biome	ecosystem	respiration
carbon cycle	habitat	succession
climate	nitrogen cycle	

1. The number of species living within an ecosystem is a measure of its

_____.

2. A somewhat regular progression of species replacement is called

_____.

3. A(n) _____ consists of a community and all the physical

aspects of its habitat, such as the soil, water, and weather.

4. Animals, plants, and other photosynthesizing organisms play important roles

in the _____ _____.

5. Ammonification is the first stage in the _____

_____.

6. A large region that is characterized by a specific kind of climate is called a

_____.

7. The many different species that live together in a habitat are called a(n)

_____.

8. The average weather conditions in an area over a long period of time is the

area's _____.

9. Calcium phosphate stored in soil and rock dissolves in water as part of the

_____ _____.

10. The exchange of oxygen and carbon dioxide between living cells and their

environment is called _____.

11. The place where a particular population of a species lives is called its

_____.

| Vocabulary Review *continued*

In the space provided, write the letter of the description that best matches the term or phrase.

_____12. community

_____13. producer

_____14. consumer

_____15. trophic level

_____16. ecosystem

_____17. succession

_____18. nitrogen cycle

_____19. climate

_____20. biome

_____21. decomposer

_____22. carbon cycle

_____23. energy pyramid

_____24. biodiversity

_____25. phosphorus cycle

_____26. respiration

_____27. habitat

a. where an organism lives

b. a triangular diagram that shows an ecosystem's loss of energy as energy passes through a food chain

c. fixes nitrogen for use by living things

d. an organism in an ecosystem that first captures energy

e. consistent weather patterns in an area over time

f. a measure based on the variety of species in an area

g. photosynthesis and respiration are two important components

h. organism that obtains energy by consuming plants or other organisms

i. root uptake and consumers eating plants are important parts of this process

j. a step in the transfer of energy through an ecosystem

k. an organism that obtains energy from organic wastes and dead bodies

l. carbon dioxide is returned to the atmosphere in this process

m. various interacting species and their physical environment

n. characterized by its climate and kinds of species

o. begins with pioneer species

p. a group of species that live in the same place and interact with one another

Name _____ Class _____ Date _____

Test Prep Pretest

In the space provided, write the letter of the term or phrase that best completes each statement or best answers each question.

_____ 1. Biodiversity is the number of species
    a. of animals living within an ecosystem.
    b. of plants and fungi living within an ecosystem.
    c. of bacteria and protists living within an ecosystem.
    d. living within an ecosystem.

_____ 2. The plants that first grow on an island formed by a volcano are part of a progression called
    a. succession.   c. competition.
    b. productivity.   d. equilibrium.

_____ 3. In the living portion of the water cycle, water
    a. is retained beneath the surface of Earth as groundwater.
    b. evaporates from the soil.
    c. evaporates from dead organisms.
    d. is taken up by the roots of plants.

Questions 4–7 refer to the figure at right.

_____ 4. The algae are
    a. decomposers.
    b. consumers.
    c. producers.
    d. herbivores.

_____ 5. The krill are
    a. decomposers.
    b. consumers.
    c. producers.
    d. herbivores.

_____ 6. This figure is called a
    a. food chain.
    b. food web.
    c. pyramid of energy.
    d. trophic level.

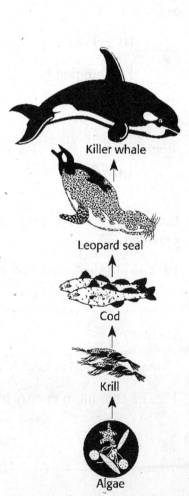

Killer whale

Leopard seal

Cod

Krill

Algae

_____ 7. The most likely reason that this figure shows only five levels is that
   a. pollution probably destroyed all of the higher levels.
   b. no other organisms are powerful enough to kill and eat the killer whale.
   c. too much energy is lost at each level to permit more levels.
   d. there is not enough energy initially present at the first level.

_____ 8. The process of succession varies depending on
   a. the plant species involved.
   b. initial environmental conditions and chance.
   c. pioneer species.
   d. competition between species.

_____ 9. The conversion of nitrate to nitrogen gas is called
   a. assimilation.   c. nitrification.
   b. ammonification.   d. denitrification.

In the space provided, write the letter of the description that best matches the term or phrase.

_____ 10. habitat

_____ 11. community

_____ 12. ecosystem

_____ 13. herbivores

_____ 14. carnivores

a. animals at the second trophic level that eat plants
b. the place where a particular population of a species lives
c. the many species that live together in a habitat
d. animals at the third trophic level that eat other animals
e. a community and all the physical aspects of its habitat

Complete each statement by writing the correct term or phrase in the space provided.

15. The physical aspects of an ecosystem, or its _____

   _____, include soil, water, and weather.

16. In a(n) _____ _____, the amount of

   energy stored at each level determines the width of each block.

17. The amount of energy that can be passed on to the third trophic level is about

   _____ percent of the amount of energy available to the

   _____ trophic level.

18. The process of combining nitrogen gas with hydrogen to form ammonia is

called _____ _____ .

19. The production of ammonia by bacteria during the decay of animal waste is

called _____ .

Read each question, and write your answer in the space provided.

20. What components are included in an ecosystem but not in a community?

_____

_____

_____

21. Why are energy pyramids never inverted?

_____

_____

_____

22. Trace the cycling of water between the atmosphere and Earth.

_____

_____

_____

_____

23. List the four stages of the nitrogen cycle.

_____

_____

_____

| Test Prep Pretest *continued*

Questions 24 and 25 refer to the figure below, which shows the carbon cycle.

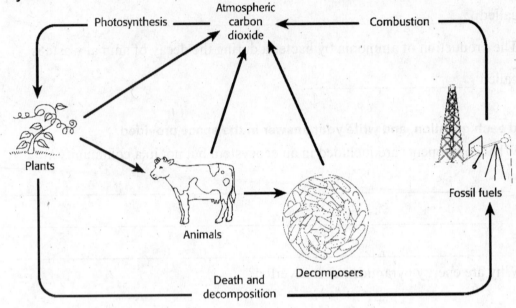

24. How do the living organisms in the figure return carbon atoms to the pool of carbon dioxide in the atmosphere and water?

_____

_____

_____

25. What process releases carbon into the atmosphere from fossil fuels?

_____

_____

Skills Worksheet

Vocabulary Review

In the space provided, write the letter of the description that best matches each term.

_____ 1. population

_____ 2. carrying capacity

_____ 3. predation

_____ 4. coevolution

_____ 5. parasitism

_____ 6. symbiosis

_____ 7. mutualism

_____ 8. commensalism

_____ 9. niche

_____ 10. fundamental niche

_____ 11. realized niche

_____ 12. competitive exclusion

_____ 13. keystone species

a. critical species in an ecosystem that affects the survival of a number of other species

b. a relationship in which both participating species benefit

c. the entire range of conditions an organism is potentially able to occupy

d. the largest population that an environment can support at any given time

e. back-and-forth evolutionary adjustments between interacting members of an ecosystem

f. two or more species living together in a close, long-term relationship

g. the unique position occupied by a species in an ecosystem

h. one organism feeds on and usually lives on or in another larger organism

i. the elimination of a competing species

j. the part of its fundamental niche that a species occupies

k. a relationship in which one species benefits and the other is neither harmed nor helped

l. a group of organisms of the same species that live together in one place at the same time

m. the act of one organism killing another organism for food

Skills Worksheet

Test Prep Pretest

In the space provided, write the letter of the term or phrase that best completes each statement or best answers each question.

_____ 1. In the exponential model of population growth, the growth rate
a. remains constant.   c. increases.
b. declines.   d. rises and falls.

_____ 2. The most important element of population growth is
a. immigration.   c. death rate.
b. emigration.   d. birthrate.

_____ 3. Most density-dependent factors that affect population growth are
a. biotic.   c. stable.
b. abiotic.   d. unimportant.

_____ 4. What form of interaction is taking place when a shark devours a seal?
a. commensalism   c. predation
b. mutualism   d. parasitism

_____ 5. When lions and hyenas fight over a dead zebra, their interaction is called
a. mutualism.   c. commensalism.
b. competition.   d. parasitism.

_____ 6. Mutualism and commensalism are two types of
a. symbiosis.   c. parasitism.
b. competition.   d. predation.

_____ 7. In the face of competition, an organism may occupy only part of its fundamental niche. That part is called its
a. biome.   c. realized niche.
b. community.   d. ecosystem.

_____ 8. The unique function an organism performs in its environment is called its
a. species.   c. niche.
b. biodiversity.   d. habitat.

_____ 9. Limited resources are the main source of
a. competition.   c. predation.
b. disease.   d. All of the above

_____ 10. The resilience of an ecosystem depends largely on which factor(s)?
a. predation   c. biodiversity
b. competition   d. All of the above

Test Prep Pretest *continued*

Complete each statement by writing the correct term or phrase in the space provided.

11. A characteristic of _____ is that they often do not kill

their prey because they depend on the prey for food and a place to live.

12. Virtually all plants contain toxic compounds that help protect the plants from

_____.

13. Rabbits that were introduced to Australia in the 1850s multiplied so rapidly

because they had no _____.

14. The entire range of conditions an organism can tolerate is its

_____ _____.

15. Back-and-forth evolutionary adjustments between interacting members of an

ecosystem are called _____.

16. When sea stars are kept out of their coastal communities, the population of

mussels in the ecosystem _____.

17. One important part of a population model is the _____

_____.

18. Density-independent factors are variables that affect a population regardless of

the population _____.

19. An important competition among plants is for the abiotic factor of

_____.

Read each question, and write your answer in the space provided.

20. Explain how the plant toxins in milkweed benefit monarch butterflies.

_____

_____

_____

_____

_____

21. Explain how predation, competition, and biodiversity are related.

_____

_____

_____

_____

22. Explain how several species of warblers that consume insects in spruce trees can occupy the same tree without competition.

_____

_____

_____

_____

23. What are two possible outcomes of competition?

_____

_____

_____

_____

| Test Prep Pretest *continued*

Questions 24–25 refer to the figure below, which shows a growth pattern of a population.

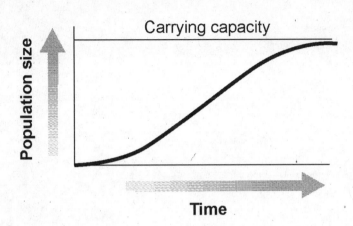

Read each question, and write your answer in the space provided.

24. What population growth model does this graph illustrate?

_____

25. Describe the changes in the line of the graph, and explain what causes the changes.

_____

_____

_____

_____

_____

_____

_____

_____

_____

Skills Worksheet

Vocabulary Review

Complete each statement by writing the correct term or phrase in the space provided.

1. When precipitation has unusually high concentrations of sulfuric or nitric acids, it

   is called _____ _____.

2. The process of recovering materials from waste or scrap is called

   _____.

3. The death of every member of a species is called _____.

4. Coal is an example of a(n) _____

   _____, which is considered a nonrenewable resource.

5. One way to educate the public about the environment is through

   _____, which involves visiting an area and learning about its

   ecosystems.

In the space provided, write the letter of the description that best matches the term or phrase.

_____ 6. global warming

_____ 7. erosion

_____ 8. deforestation

_____ 9. biodiversity

_____ 10. greenhouse effect

a. the process of clearing land for timber or to make farmland

b. a gradual increase in average temperatures on Earth

c. the warming of the atmosphere due to the absorption and reradiation of heat

d. the wearing away of Earth's surface by wind, gravity, or water

e. the variety of organisms in an area

Skills Worksheet

Test Prep Pretest

In the space provided, write the letter of the term or phrase that best completes each statement or best answers each question.

_____ 1. When sulfur dioxide is released into the atmosphere, the result often is
    a. ozone.
    b. CFCs.
    c. acid rain.
    d. ultraviolet radiation.

_____ 2. Global levels of carbon dioxide are
    a. increasing.
    b. remaining constant.
    c. decreasing.
    d. too low to be measured accurately.

_____ 3. All of the following are considered nonrenewable resources *except*
    a. oil.
    b. wood.
    c. coal.
    d. natural gas.

_____ 4. Temperatures on Earth are kept stable due to
    a. global warming.
    b. acid rain.
    c. ozone buildup.
    d. the greenhouse effect.

_____ 5. The increase in global temperatures over the last 45 years is associated with
    a. decreased CFCs.
    b. increased carbon dioxide.
    c. acid rain.
    d. the greenhouse effect.

_____ 6. Efforts to reduce pollution include all of the following *except*
    a. restrictions on the use of DDT.
    b. use of scrubbers in industry.
    c. limits on CFC use.
    d. the closing of all coal-burning facilities.

_____ 7. Cleaning up damaged habitats is a function of which technique for solving environmental problems?
    a. restoration
    b. conservation
    c. reduction
    d. ecotourism

| Test Prep Pretest *continued*

Complete each statement by writing the correct term or phrase in the space provided.

8. The ozone layer has been damaged by the use of _____, used as coolants in refrigerators.

9. The insulating effect of various gases in Earth's atmosphere is known as the _____ _____.

10. The increase in global temperatures is called _____ _____.

11. Examples of chemical pollutants released into the global ecosystem by the agriculture industry are _____ and _____.

12. Fossil fuels are formed when _____ are buried by layers of sediments that cause intense heat and pressure.

13. Fertilizer runoff may cause _____ _____ that deplete the amount of _____ in the water.

14. Washing cars and watering lawns less often and using efficient faucets are ways to _____ water usage.

15. The installation of _____ on factory smokestacks has reduced sulfur dioxide and carbon monoxide emissions by 30 percent.

| Test Prep Pretest *continued*

Read each question, and write your answer in the space provided.

16. How does the presence of the ozone layer affect life on Earth?

_____

_____

_____

17. Explain the relationship between the greenhouse effect and global warming.

_____

_____

_____

_____

_____

18. Explain how runoff from farms and golf courses can affect organisms in nearby streams.

_____

_____

_____

_____

19. Describe four techniques for conserving soil.

_____

_____

_____

_____

20. Explain the relationship between habitat destruction and loss of biodiversity.

_____

_____

_____

_____

Skills Worksheet

Vocabulary Review

In the space provided, write the letter of the description that best matches each term.

_____ 1. colonial organism

_____ 2. cell membrane

_____ 3. cytoplasm

_____ 4. ribosome

_____ 5. endoplasmic reticulum

_____ 6. tissue

_____ 7. organ

_____ 8. organ system

_____ 9. prokaryote

_____ 10. eukaryote

_____ 11. nucleus

_____ 12. organelle

_____ 13. vesicle

_____ 14. vacuole

_____ 15. Golgi apparatus

_____ 16. mitochondrion

_____ 17. chloroplast

_____ 18. flagellum

a. organelle that contains the DNA of a eukaryotic cell

b. small sac that contains materials

c. collection of organs that carry out a major body function

d. organelle that uses energy from organic compounds to make ATP

e. distinct group of cells with a similar structure and function

f. group of cells that are permanently associated but do not communicate with one another

g. long, threadlike structure that rotates quickly to aid cell movement

h. cell structure that carries out a specific, specialized function in a eukaryotic cell

i. organelle that helps package materials to be sent out of the cell

j. tissues organized into a specialized structure with a specific function

k. layer that forms the boundary between the inside and the outside of a cell

l. fluid in a cell and almost all of the structures suspended in the fluid

m. sac filled with fluid in a plant cell

n. cell structure on which proteins are made

o. organism that has no nucleus and lacks a variety of organelles

p. system of membranes that helps move substances through a cell

q. organism made up of cells that have a nucleus and membrane-bound organelles

r. organelle that uses light energy to make sugar from carbon dioxide and water

Skills Worksheet

Test Prep Pretest

In the space provided, write the letter of the term or phrase that best completes each statement or best answers each question.

_____ 1. The surface area-to-volume ratio of a small cell is
a. greater than that of a larger cell.
b. less than that of a larger cell.
c. equal to that of a larger cell.
d. not affected by the cell's size.

_____ 2. In prokaryotic cells, the genetic material is found in
a. the DNA and RNA.   c. the nucleus.
b. the nucleolus.   d. a single loop.

_____ 3. In eukaryotic cells, mitochondria
a. transport materials.   c. produce ATP.
b. make proteins.   d. control cell division.

_____ 4. Which cell structures do all bacteria and plants have in common?
a. chloroplasts   c. a cell wall
b. pili   d. Both (a) and (c)

_____ 5. Which of these are always unicellular?
a. prokaryotes   c. protists
b. eukaryotes   d. flagella

_____ 6. Most animals and plants have groups of cells with a similar structure and function that are organized into
a. organ systems.   c. nerves and muscles.
b. tissues.   d. All of the above

Questions 7 and 8 refer to the figure at right.

_____ 7. The cell in the figure is a
a. prokaryotic cell.
b. eukaryotic cell.
c. plant cell.
d. Both (b) and (c)

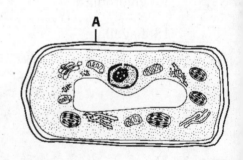

_____ 8. The structure labeled *A*
a. supports the cell.
b. protects the cell.
c. surrounds the cell membrane.
d. All of the above

| Test Prep Pretest *continued*

Complete each statement by writing the correct term or phrase in the space provided.

9. Scientists first discovered cells by using a(n) _____.

10. A cell's boundary is called the _____

    _____.

11. _____ are cell structures common to both prokaryotes and

    eukaryotes on which proteins are made.

12. Eukaryotes differ from prokaryotes in that only eukaryotic cells have a(n)

    _____ and membrane-bound _____.

13. The nucleus has a double membrane, called the nuclear envelope, that helps

    protect a cell's _____ from becoming damaged or lost.

14. In plant cells, rigidity is provided by a large, membrane-bound sac called the

    _____ _____.

15. When a cell makes proteins that are to be transported outside the cell, the

    proteins are packaged in the _____

    _____, modified and repackaged in the

    _____ _____, and then transported to

    the cell membrane.

16. Vesicles which contain enzymes that break down large molecules are called

    _____.

17. The _____ is a network of protein fibers that supports a

    cell and aids in its movement.

18. Organelles that use light energy to make sugar from water and carbon dioxide

    are called _____.

19. A(n) _____ is made up of different kinds of tissues

    arranged together to perform a specific function.

20. A collection of identical cells that live together as a group, although individuals

can survive on their own, is called a(n) _____

_____ .

21. Cells in a(n) _____ organism cannot survive on their own.

Questions 22–28 refer to the figure below.

22. The structure labeled *A* is the _____ _____ .

23. The organelle labeled *B* is the _____ _____ .

24. The structure labeled *C* is the _____ _____ .

25. The structure labeled *D* is the _____ _____ .

26. The organelle labeled *E* is the _____ _____ .

27. The organelle labeled *F* is a(n) _____ .

28. The organelle labeled *G* is a(n) _____ .

Read each question, and write your answer in the space provided.

29. List the three parts of the cell theory.

_____

_____

_____

_____

Name _____ Class _____ Date _____

30. List the primary differences between prokaryotes and eukaryotes.

_____

_____

_____

_____

_____

_____

Skills Worksheet

Vocabulary Review

In the space provided, write the letter of the description that best matches each term.

_____ 1. concentration gradient

_____ 2. equilibrium

_____ 3. diffusion

_____ 4. osmosis

_____ 5. phospholipid

_____ 6. carrier protein

_____ 7. receptor protein

_____ 8. lipid bilayer

_____ 9. signal

_____ 10. sodium-potassium pump

_____ 11. second messenger

a. the movement of a substance from a region where its concentration is higher to a region where its concentration is lower

b. transports specific substances across a cell membrane

c. binds to a signal molecule, enabling the cell to respond to the signal molecule

d. made of a phosphate group and two fatty acids

e. acts as a signal molecule in a cell's cytoplasm

f. the state in which the distribution of a substance is even throughout a region

g. the difference in the concentration of a substance across a distance

h. type of carrier protein that uses active transport to take sodium ions out of the cell and bring potassium ions into the cell

i. a double layer of phospholipids that is the foundation of a biological membrane

j. the movement of water through a selectively permeable membrane from a more dilute solution to a more concentrated solution

k. anything that carries information between cells, serving to direct or warn

Name _____ Class _____ Date _____

Skills Worksheet

Test Prep Pretest

In the space provided, write the letter of the description that best matches each term.

_____ 1. phospholipid

_____ 2. cell-surface marker

_____ 3. receptor protein

_____ 4. channel protein

_____ 5. carrier protein

_____ 6. diffusion

_____ 7. osmosis

_____ 8. equilibrium

_____ 9. signal

_____10. second messenger

a. the movement of water from a region of higher concentration to a region of lower concentration, passing through a selectively permeable membrane

b. a substance, located in the cell membrane and made of amino acids, that binds with specific molecules, causing a change in the cell

c. most often a molecule that serves to carry information between cells

d. a substance, located in the cell membrane and made of amino acids, which other substances can pass through to cross the cell membrane

e. the state in which the distribution of a substance is even throughout a region

f. a substance, located in the cell membrane and made of amino acids, that moves other substances across the cell membrane

g. a substance, located in the cell membrane and made of amino acids and sugars, that aids in the identification of cell type

h. the movement of a substance from a region of higher concentration to a region of lower concentration

i. a substance made of a phosphate group and two fatty acids

j. a substance, generated when a signal molecule binds with a receptor protein, that serves as a signal inside the cell

In the space provided, write the letter of the term or phrase that best completes each statement or best answers each question.

_____11. Which of the following best describes the cell membrane?
    a. waterproof layer of sugars connected to surface proteins
    b. single layer of amino acids
    c. double layer of phospholipids
    d. thick layer of glycoproteins

_____12. What is the difference between passive transport and active transport?
    a. Passive transport requires energy, and active transport does not.
    b. Active transport requires energy, and passive transport does not.
    c. Passive transport uses carrier proteins, and active transport does not.
    d. Active transport uses carrier proteins, and passive transport does not.

_____13. The discharging of materials to the outside of a cell using vesicles is called
    a. exocytosis.   c. passive transport.
    b. endocytosis.   d. channeling proteins.

_____14. When a receptor protein in a cell membrane acts as an enzyme, the receptor protein
    a. changes its shape to allow the signal molecule to enter the cell.
    b. triggers a chemical reaction in the cell.
    c. activates a second messenger that acts as a signal molecule within the cell.
    d. changes the permeability of the cell membrane.

_____15. When a particle moves across a cell membrane from an area of low concentration to an area of higher concentration, the cell is using
    a. diffusion.   c. osmosis.
    b. facilitated diffusion.   d. active transport.

_____16. Receptor proteins have binding sites, each with a unique shape because
    a. the outer amino acids fold in a complex pattern.
    b. the inner amino acids arrange themselves in a dense ball.
    c. vesicles change the shape of the cell membrane.
    d. these proteins can respond to light from the environment.

_____17. Which of these are the two categories of transport proteins?
    a. receptor proteins and glycoproteins
    b. glycoproteins and channel proteins
    c. channel proteins and carrier proteins
    d. carrier proteins and receptor proteins

| Test Prep Pretest *continued*

Question 18 refers to the figure at right.

18. The process shown in the figure is

_____.

Complete each statement by writing the correct term or phase in the space provided.

19. The head of a phospholipid is

_____, so it is attracted to water. The tails are

_____, so they are repelled by water.

20. Membrane proteins remain stable in a cell membrane because the nonpolar

amino acids in each protein are attracted to the _____ of

the lipid bilayer, while the polar amino acids in each protein are attracted to

the _____ on either side of the cell membrane. This

creates a tension in all membrane proteins that holds them in place.

21. When a substance moves from an area of low concentration to an area of

higher concentration, the substance moves _____ its

_____ concentration gradient.

22. Plant cells are healthiest in a(n) _____ solution because

they swell with water, helping to give the plant support.

23. Two cells communicate when a(n) _____ sent by one cell

binds with a(n) _____ _____ in the

membrane of another cell, causing the latter to change shape. This relays

information into the second cell's cytoplasm.

Read each question, and write your answer in the space provided.

24. How does the cell membrane help a cell maintain homeostasis?

_____

_____

_____

25. How does facilitated diffusion differ from simple diffusion? Give examples of each.

_____

_____

_____

_____

_____

_____

26. Describe the purpose of the sodium-potassium pump, and explain how it works.

_____

_____

_____

_____

_____

27. Why is osmosis important for cells?

_____

_____

28. How does a cell consume a food particle that is too large to pass through a channel protein?

_____

_____

_____

29. What are three different ways that a cell can respond to a signal?

_____

_____

_____

Skills Worksheet

Vocabulary Review

Write the correct term from the list below in the space next to its definition.

aerobic	cellular respiration	Krebs cycle
anaerobic	chlorophyll	photosynthesis
ATP	electron transport chain	pigment
ATP synthase	fermentation	thylakoid
Calvin cycle	glycolysis	

_____ 1. the process some organisms are able to use by which they convert light energy to chemical energy

_____ 2. the main method photosynthesizing organisms use for carbon dioxide fixation

_____ 3. a set of chemical reactions that break down pyruvate, producing electron carriers for an electron transport chain that powers ATP production

_____ 4. the process cells use to produce energy from carbohydrates

_____ 5. a substance that absorbs some wavelengths of light and reflects others, giving something its color

_____ 6. the green substance that absorbs light and provides energy for photosynthesis

_____ 7. disc-shaped sacs in chloroplasts in which photosynthesis takes place

_____ 8. the series of molecules in the inner membranes of chloroplasts and mitochondria down which excited electrons pass, releasing energy for ATP production

_____ 9. process by which NAD^+ is recycled under anaerobic conditions in order to continue the break down of carbohydrates to supply energy for producing ATP

_____ 10. adenosine triphosphate, a substance that stores and releases energy for most cell processes

_____ 11. describes a process that requires oxygen

_____ 12. describes a process that does not require oxygen

_____ 13. the process by which glucose is broken down into pyruvate in the absence of oxygen, producing a small amount of ATP

_____ 14. the enzyme that aids in the production of adenosine triphosphate and which also acts as a carrier protein for hydrogen ions in active transport across a membrane

Skills Worksheet

Test Prep Pretest

In the space provided, write the letter of the term or phrase that best completes each statement or best answers each question.

_____ 1. Which of the following correctly sequences the flow of energy through an ecosystem?
   a. bacteria, water, algae, fish
   b. bacteria, sun, grass, deer
   c. sun, grass, rabbit, fox
   d. algae, sun, small fish, shark

_____ 2. What is the purpose of cellular respiration?
   a. to store carbohydrates
   b. to produce energy from carbohydrates
   c. to produce oxygen
   d. to store oxygen in water

_____ 3. What is the main way cells get energy from ATP?
   a. by using water to release energy from the molecule
   b. by breaking the single phosphate bond in the molecule
   c. by breaking one of the two phosphate bonds in the molecule
   d. by breaking one of the three phosphate bonds in the molecule

_____ 4. ATP synthase gets the energy it needs to make ATP directly from
   a. hydrogen ions diffusing through the channel in the protein.
   b. hydrogen ions it pumps out of the cell, across the cell membrane.
   c. electrons in pigments that have absorbed sunlight.
   d. electrons that have bonded to carbohydrate molecules.

_____ 5. Electron transport chains are a series of molecules
   a. on the inner membrane of some organelles that accept excited electrons and use their energy to move H^+ ions across the membrane.
   b. on the outer membrane of some organelles that accept H^+ ions and use their energy to move electrons across the membrane.
   c. on the inside of some cell membranes that accept H^+ ions and use their energy to move protons out of the cell.
   d. on the outside of some cell membranes that accept excited elections and use their energy to move H^+ ions into the cell.

_____ 6. How do plants get energy from light?
   a. Light excites hydrogen ions in the outer membrane of chloroplasts.
   b. Light excites electrons in a special chlorophyll molecule.
   c. Light excites ATP synthase in the membrane of plant cells.
   d. Light excites electron carriers in carotenoids.

_____ 7. Which do plants need to complete making sugars via photosynthesis?
   a. glucose   b. oxygen   c. alcohol   d. carbon dioxide

_____ 8. Which of the following environmental factors does *not* directly
influence the rate of photosynthesis?
a. light intensity
b. oxygen concentration
c. carbon dioxide concentration
d. temperature

_____ 9. In glycolysis,
a. glucose is produced.
b. aerobic processes produce energy-storing sugars.
c. one molecule of ATP and one molecule of NADPH are produced.
d. one molecule of pyruvate and four molecules of ATP are produced.

_____ 10. Which of these occurs during the Calvin cycle?
a. An animal cell produces a net total of two molecules of ATP.
b. An animal cell produces up to a net total of 36 molecules of ATP.
c. A plant cell produces one energy-storing sugar molecule.
d. A plant cell produces up to 36 energy-storing sugar molecules.

_____ 11. Which of these are end products of the Krebs cycle?
a. ATP, NADH, and FADH$_2$
b. ATP and oxygen
c. ATP and pyruvate
d. ATP and energy-storing starch

_____ 12. Which of the following is *never* part of cellular respiration?
a. an electron transport chain
b. glycolysis
c. the Krebs cycle
d. the Calvin cycle

_____ 13. The most efficient form of cellular respiration requires
a. carbon dioxide as a source for making energy-storing molecules.
b. water as a source of excited electrons in electron transport chains.
c. oxygen as an electron acceptor so electron carriers can be recycled.
d. lactic acid as an electron acceptor so electron carriers can be
recycled.

Question 14 refers to the figure below, which shows a chloroplast.

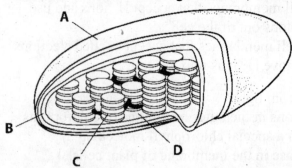

_____ 14. The reactions of the electron transport chains occur in the structure
labeled
a. *A.*
b. *B.*
c. *C.*
d. *D.*

_____15. Which substances are involved in two different types of fermentation?
    a. ethanol and lactic acid   c. glucose and oxygen
    b. glucose and ethanol   d. lactic acid and glucose

_____16. What is the net maximum number of ATP molecules that can be
    produced through cellular respiration?
    a. 41   c. 34
    b. 38   d. 2

Complete each statement by writing the correct term or phrase in the space provided.

17. During photosynthesis, organisms called _____ convert

_____ energy to _____ energy.

18. Cells gradually release energy in chemical reactions that are assisted by

catalysts called _____.

19. Light-absorbing _____ are located in the membranes of

flat sacs called _____, which are stacked inside

_____.

20. During the light reactions of photosynthesis, one _____

_____ _____ uses energy from

excited electrons to make _____ the other uses energy from excited

electrons to make _____. These molecules are used in the

_____ _____ of photosynthesis.

21. Some organisms use the process of fermentation to resupply electron acceptors

for _____, so that ATP can be produced in the absence of

_____.

22. Aerobic respiration occurs in the _____ of eukaryotic cells.

Read each question, and write your answer in the space provided.

23. Explain how organisms participate in Earth's carbon cycle through metabolic processes.

_____

_____

_____

_____

_____

24. Explain how the metabolism of organisms that are not autotrophic differs from that of autotrophs.

_____

_____

_____

_____

25. What is ATP and why is it considered a form of "energy currency" for a cell?

_____

_____

_____

_____

26. Briefly explain how ATP is produced by electron transport chains.

_____

_____

_____

_____

_____

_____

_____

Skills Worksheet

Vocabulary Review

In the space provided, write the letter of the term or phrase that best completes each statement or best answers each question.

_____ 1. A segment of DNA that codes for RNA and protein is a
    a. chromosome.
    b. chromatid.
    c. gene.
    d. centromere.

_____ 2. The structure in a cell that is made up of the cell's genetic material is a
    a. centriole.
    b. chromosome.
    c. centrosome.
    d. histone.

_____ 3. Which of these is a substance consisting of DNA and protein?
    a. chromatin
    b. centriole
    c. histone
    d. centrosome

_____ 4. The repeated sequence of growth and division during the life of a cell is called
    a. the cell cycle.
    b. the life cycle.
    c. mitosis.
    d. cytokinesis.

_____ 5. The first three phases of the life cycle of a cell are called
    a. anaphase.
    b. interphase.
    c. the first gap phase.
    d. the synthesis phase.

_____ 6. What is the process during which the nucleus of a cell is divided into two nuclei?
    a. the cell cycle
    b. nucleosome
    c. mitosis
    d. cytokinesis

_____ 7. The phase of cell division when the cytoplasm is divided is called
    a. the first gap phase.
    b. the second gap phase.
    c. the synthesis phase.
    d. cytokinesis.

_____ 8. During cell division, each single long strand of DNA becomes fully condensed in the form of a
    a. centrosome.
    b. chromosome.
    c. chromatin.
    d. chromatid.

_____ 9. A type of protein found in eukaryotic chromosomes but not prokaryotic chromosomes is
    a. centrosome.
    b. nucleosome.
    c. spindle fiber.
    d. histone.

_____10. During cell division, sister chromatids are separated at the
        a. centromere.   c. centrosome.
        b. nucleosome.   d. chromosome.

_____11. Which of these is a network of microtubules that forms during mitosis
to pull chromatids to opposite ends of a cell?
        a. histone   c. spindle
        b. chromatin   d. centromere

_____12. In eurkaryotes, a structural unit made up of DNA wound around a
center of histone proteins is called a
        a. chromatid.   c. centrosome.
        b. nucleosome.   d. looped domain.

_____13. The structure that directs chromosome movement during mitosis and
aids in the formation of the microtubule scaffolding that pulls on the
chromosomes is the
        a. centrosome.   c. spindle.
        b. nucleosome.   d. cytokinesis.

Complete each statement by writing the correct term in the space provided.

14. _____ is a group of severe and sometimes fatal diseases

caused by uncontrolled cell growth.

15. A mass of defective cells that divide very rapidly is called a(n)

_____.

Skills Worksheet

Test Prep Pretest

In the space provided, write the letter of the term or phrase that best completes each statement or best answers each question.

_____ 1. Large cells are difficult to maintain. How do cells overcome this problem?
   a. They double their DNA.   c. They take in more nutrients.
   b. They coil DNA around proteins. d. They undergo cell division.

_____ 2. How do chromosomes of eukaryotes compare with chromosomes of prokaryotes?
   a. Eukaryotes have a single chromosome, whereas prokaryotes have a number of chromosomes.
   b. Eukaryotes have chromosomes in the form of a ring, whereas prokaryotes have condensed chromatin in their chromosomes.
   c. Eukaryotes have chromosomes made of DNA and proteins in a condensed form, whereas prokaryotes have a twisted loop of DNA.
   d. Both (a) and (b)

_____ 3. DNA is coiled in chromosomes so it can
   a. be packed into a small space.
   b. wind around the essential proteins.
   c. can code for RNA and proteins.
   d. form a centrosome.

_____ 4. What do all cells need to do before they begin to divide?
   a. form a daughter cell   c. uncoil their DNA
   b. form a cell plate   d. copy their DNA

_____ 5. In which order do the four stages of mitosis occur?
   a. anaphase, interphase, prophase, and telophase
   b. prophase, anaphase, metaphase, and telophase
   c. interphase, prophase, anaphase, and telophase
   d. prophase, metaphase, anaphase, and telophase

_____ 6. During which phase of mitosis do chromatids line up along the equator of the dividing cell?
   a. anaphase   c. interphase
   b. metaphase   d. prophase

_____ 7. Which of these structures is found only in a dividing animal cell and not in any other type of dividing cell?
   a. centrioles   c. a spindle
   b. centrosomes   d. sister chromatids

_____ 8. How does cytokinesis in animal cells differ from cytokinesis in plant cells?
  a. In animal cells, the loop of DNA attaches to the cell membrane, whereas in plant cells it does not.
  b. In animal cells, the formation of the cell membrane involves vesicles, whereas in plant cells it does not.
  c. In animal cells, protein threads pinch the dividing cell in half, whereas in plant cells a cell plate forms down the middle of the dividing cell.
  d. Both (a) and (b)

_____ 9. What slows the rapid cell division of cells healing a cut in the skin?
  a. the need to copy DNA
  b. the need to renew the tissues
  c. contact with other skin cells
  d. contact with vesicles in the middle of the cell

_____ 10. How does a cell ensure that no mistakes occur in the DNA when the cell is dividing?
  a. Environmental signals influence the cell cycle.
  b. Protein signals from nearby cells affect a dividing cell.
  c. There is a checkpoint during mitosis.
  d. There is a checkpoint before mitosis begins.

On the line before the term, rank the term to show the level of packaging from least condensed to most condensed. On the line after each term, define or describe the term.

_____ 11. histone core _____

_____ 12. DNA _____

_____ 13. looped domain _____

_____ 14. nucleosome cord _____

_____ 15. one histone _____

_____ 16. chromatid _____

_____ 17. nucleosome _____

Complete each statement by writing the correct term in the space provided.

18. A(n) _____ is a segment of DNA that codes for RNA and

protein.

19. The material that makes up chromosomes in eukaryotic cells and which is formed of DNA and protein is called _____.

20. The organelle that is the center of dynamic activity in a dividing cell is the

_____.

21. The network of microtubules that pulls chromatids to the poles as a cell is dividing is called the _____.

22. Sister chromatids attach to each other in the region called the

_____.

23. _____ is a group of diseases caused by uncontrolled cell growth.

Questions 24–31 refer to the sequence below.

$$G_1 \longrightarrow S \longrightarrow G_2 \longrightarrow M \longrightarrow C$$

24. The sequence above represents the _____

_____.

25. The S in the sequence represents the phase in which _____

occurs.

26. Phases G_1, S, and G_2 in the sequence above are collectively called

_____.

27. During _____, a cell nucleus divides into two separate nuclei.

28. Two daughter cells form during _____.

29. The checkpoint in which the cell checks to make sure that chromatids are correctly attached to the spindle occurs between the _____ phase and the _____ phase.

30. The checkpoint in which the cell checks to make sure the cell is healthy and

    large enough and that surrounding conditions are favorable occurs between the

    _____ phase and the _____ phase.

31. What happens after the C phase?

    _____

    _____

Read each question, and write your answer in the space provided.

32. Why does the body of a multicellular organism grow larger through cell
    division rather than by simply growing larger cells? Give two reasons.

    _____

    _____

    _____

    _____

    _____

    _____

33. What is a tumor? How does a benign tumor differ from a malignant tumor?

    _____

    _____

    _____

    _____

    _____

    _____

    _____

Skills Worksheet

Vocabulary Review

In the space provided, write the letter of the term or phrase that best completes each statement or best answers each question.

_____ 1. An organism's reproductive cells, such as sperm or egg cells, are called
        a. genes.   c. gametes.
        b. chromosomes.   d. zygotes.

_____ 2. Chromosomes that are similar in size, shape, and genetic content are called which of the following?
        a. homologous chromosomes   c. diploid
        b. haploid   d. ovum

_____ 3. When a cell contains two sets of chromosomes, it is said to be
        a. haploid.   c. diploid.
        b. binary.   d. budding.

_____ 4. When a cell contains one set of chromosomes, it is said to be
        a. haploid.   c. diploid.
        b. crossing-over.   d. homologous.

_____ 5. A type of cell division that halves the number of chromosomes is known as
        a. anaphase.   c. mitosis.
        b. meiosis.   d. gametophyte.

_____ 6. What process occurs during prophase I of meiosis?
        a. cytokinesis   c. crossing-over
        b. random fertilization   d. chromosome

_____ 7. The union of the gametes during fertilization leads to the production of a(n)
        a. alternation of generations.   c. organism.
        b. zygote.   d. chromosome.

_____ 8. The random distribution of homologous chromosomes in the formation of the gametes is
        a. independent assortment.   c. crossing-over.
        b. zygote.   d. chromosome.

_____ 9. What gamete is produced in quantities of four cells during meiosis?
        a. asexual
        b. ovum
        c. sporophyte
        d. sperm

_____ 10. What gamete is produced as one large cell and three smaller cells during meiosis?
        a. asexual
        b. ova
        c. sporophyte
        d. sperm

_____ 11. Diploid is an example of a
        a. life cycle.
        b. cell cycle.
        c. haploid cycle.
        d. sperm cell.

Name _____ Class _____ Date _____

Test Prep Pretest

Complete each statement by writing the correct term or phrase in the space provided.

1. Asexual reproduction limits _____ diversity.

2. Spermatogenesis produces _____ sperm cells.

3. Asexual reproduction methods include _____,

   fragmentation, parthenogenesis, and _____

   _____.

4. In the haploid life cycle, gametes are produced by _____, and the

   zygote is produced by _____.

5. When corresponding portions of chromatids on two homologous chromosomes

   change places, _____-_____ has

   occurred.

6. Only one ovum is produced by _____.

7. In plants that have alternation of generations, the haploid

   _____ produces the gametes.

8. Increased genetic variation often helps organisms withstand changes in the

   _____.

9. Meiosis in plants often produces _____, haploid cells that

   later lead to the production of gametes.

10. Crossing-over and _____ _____ produce

    genetic diversity.

11. The 22 pairs of chromosomes in human somatic cells that are the same in

    males and females are called _____.

12. The human chromosomes that determine an individual's sex are called the

    _____ _____.

Questions 11–14 refer to the figure below.

A   B   C   D

13. The process shown above is called _____.

14. In the process shown above, label *A* refers to _____.

15. In the process shown above, label *B* refers to _____ and

    _____.

16. In the process shown above, label *C* refers to _____.

Read each question, and write your answer in the space provided.

17. Describe the similarities and differences between the formation of male and
    female gametes.

    _____

    _____

    _____

    _____

    _____

18. Identify and describe four types of asexual reproduction.

    _____

    _____

    _____

    _____

    _____

    _____

    _____

19. What is the difference between anaphase I and anaphase II? Why is the difference significant?

_____

_____

_____

_____

_____

20. Describe the haploid and diploid life cycles.

_____

_____

_____

_____

_____

21. Describe the advantages of sexual reproduction.

_____

_____

_____

_____

22. Explain the difference in the number of chromosomes between a frog somatic cell and a frog egg cell.

_____

_____

_____

_____

Skills Worksheet

Vocabulary Review

In the space provided, write the letter of the description that best matches the term or phrase.

_____ 1. character

_____ 2. trait

_____ 3. hybrid

_____ 4. generation

_____ 5. allele

_____ 6. dominant

_____ 7. recessive

_____ 8. homozygous

_____ 9. heterozygous

_____ 10. genotype

_____ 11. phenotype

a. when the two alleles of a particular gene are different

b. the allele that is not expressed when the dominant corresponding allele is present

c. an inherited feature or characteristic

d. the offspring of a cross between parents that have contrasting traits

e. the allele that is fully expressed by itself

f. a detectable trait or traits that result from the genotype

g. a form of a character

h. when the two alleles of a particular gene are the same

i. a version of a gene

j. the set of alleles that an individual has

Write the correct term from the list below in the space next to its definition.

codominance   pedigree   Punnett square
genetic disorder   polygenic character
linked   probability

_____ 12. a model that predicts the likely outcomes of a genetic cross

_____ 13. the likelihood that a specific event will occur

_____ 14. a family history that shows how a trait is inherited

_____ 15. when several genes influence a character

_____ 16. a condition in which both alleles for a gene are fully expressed at the same time

_____ 17. an inherited disease or disorder

_____ 18. genes located close together on the same chromosome

Skills Worksheet

Test Prep Pretest

In the space provided, write the letter of the term or phrase that best completes each statement or best answers each question.

_____ 1. *Pisum sativum*, the garden pea, is a good subject to use in studying heredity for all of the following reasons *except*
   a. Several varieties of *Pisum sativum* are available that differ in easily distinguishable traits.
   b. *Pisum sativum* is a small, easy-to-grow plant.
   c. *Pisum sativum* matures quickly and produces a large number of offspring.
   d. A *Pisum sativum* plant with male reproductive parts must cross-pollinate with a plant having female reproductive parts for reproduction to take place.

_____ 2. Step 1 of Mendel's garden pea experiment, allowing each variety of garden pea to self-pollinate for several generations, produced the
   a. F_1 generation.   c. P generation.
   b. F_2 generation.   d. P_2 generation.

_____ 3. In the F_2 generation in Mendel's experiments, the ratio of dominant to recessive phenotypes was
   a. 1:3.   c. 2:1.
   b. 1:2.   d. 3:1.

_____ 4. The trait that was expressed in the F_1 generation in Mendel's experiment is considered
   a. recessive.   c. second filial.
   b. dominant.   d. parental.

_____ 5. Mendel's law of segregation states that
   a. pairs of alleles are dependent on one another when separation occurs during gamete formation.
   b. pairs of alleles separate independently of one another after gamete formation.
   c. each pair of alleles remains together when gametes are formed.
   d. the two alleles for a gene separate when gametes are formed.

_____ 6. A series of genetic crosses results in 787 long-stemmed plants and 277 short-stemmed plants. The probability that you will obtain short-stemmed plants if you repeat this experiment is

   a. $\dfrac{277}{1,064}$.   c. $\dfrac{787}{277}$.

   b. $\dfrac{277}{787}$.   d. $\dfrac{787}{1,064}$.

| Test Prep Pretest *continued*

_____ 7. Crossing a snapdragon that has red flowers with one that has white flowers produces a snapdragon that has pink flowers. The trait for flower color exhibits
   a. multiple alleles.   c. incomplete dominance.
   b. complete dominance.   d. codominance.

_____ 8. Which of the following is *not* considered a simple Mendelian inheritance pattern?
   a. recessive/dominant traits   c. polygenic inheritance
   b. law of segregation   d. law of independent assortment

_____ 9. On which of the following chromosomes would a sex-linked gene more often be found in humans?
   a. X   c. O
   b. Y   d. YO

_____ 10. The human blood groups are an example of
   a. homozygous alleles.   c. incomplete dominance.
   b. codominance.   d. Both (a) and (c)

Questions 11 and 12 refer to the figure at right, which represents a monohybrid cross between two individuals that are heterozygous for a trait.

_____ 11. If the resulting phenotypic ratio is 3:1, the missing parental allele is
   a. *d*.   c. *Dd*.
   b. *D*.   d. *DD*.

	D	**d**
D	**DD**	**Dd**
D__	**D__**	**d__**

_____ 12. The two unknown genotypes in the offspring are
   a. *DD* and *dd*.   c. *dd* and *DD*.
   b. *Dd* and *Dd*.   d. *Dd* and *dd*.

_____ 13. Which of the following summarizes one of Mendel's major hypotheses developed from his studies of garden peas?
   a. All of an individual's alleles make up its genotype.
   b. Traits that are intermediate between two parents are caused by genes that are incompletely dominant.
   c. There are alternative versions of genes, which are now called alleles.
   d. When two dominant alleles are expressed together, they are called codominant.

_____ 14. Which of the following is an example of a testcross?
   a. *YY* × *YY*
   b. *YY* × *yy*
   c. *Yy* × *Yy*
   d. All of the above

Test Prep Pretest *continued*

_____15. What is the relationship between genotype and phenotype?
   a. Phenotype determines a genotype.
   b. Genotype produces a phenotype.
   c. Genotype and phenotype give rise to alleles.
   d. None of the above

Question 16 refers to the figure below, which shows the inheritance of sickle cell anemia in a family.

_____16. Which of the following is true based on the information provided in the pedigree?
   a. Both parents have sickle cell anemia.
   b. Both parents carry an allele for sickle cell anemia.
   c. Sickle cell anemia is caused by a dominant allele.
   d. All three children are carriers of a defective gene that causes sickle cell anemia.

Complete each statement by writing the correct term or phrase in the space provided.

17. The investigator whose studies formed the basis of modern genetics is

   _____ .

18. The _____ , or detectable trait, of an individual is

   determined by the alleles that code for traits. The set of alleles that an

   individual has is called its _____ .

19. A cross between a pea plant that is true-breeding for green pod color and one

   that is true-breeding for yellow pod color is an example of a(n)

   _____ cross.

20. Characteristics such as eye color, height, weight, and hair and skin color are

    examples of _____ _____ because

    several genes act together to influence a character.

21. Genes that are close together on the same chromosome are said to be

    _____.

Read each question, and write your answer in the space provided.

22. What approximate ratio of plants expressing contrasting traits did Mendel
    calculate in his F_2 generation of garden peas? What steps did he take to
    calculate this ratio?

    _____

    _____

    _____

    _____

23. Name Mendel's two major laws of heredity.

    _____

    _____

24. Give an example of how the environment might influence gene expression.

    _____

    _____

    _____

25. Describe the inheritance of sex-linked genes.

    _____

    _____

    _____

Skills Worksheet

Vocabulary Review

In the space provided, write the letter of the description that best matches the term or phrase.

_____ 1. DNA

_____ 2. nucleotide

_____ 3. purines

_____ 4. genes

_____ 5. pyrimidines

_____ 6. DNA helicase

_____ 7. DNA polymerase

_____ 8. DNA replication

a. represented by adenine and guanine

b. enzyme that separates the DNA helix by breaking the hydrogen bonds that link the nitrogenous bases

c. instructions for inherited traits

d. the process by which DNA is copied

e. represented by thymine and cytosine

f. consists of a phosphate, a five-carbon sugar, and a nitrogenous base

g. enzyme that adds nucleotides to exposed nitrogenous bases

h. name given for deoxyribonucleic acid

Write the correct term from the list below in the space next to its definition.

| codon | RNA | translation |
| gene expression | transcription | |

_____ 9. the process in which RNA is made from the information in DNA

_____ 10. includes transcription and translation

_____ 11. a three-nucleotide sequence that encodes an amino acid or a start/stop signal

_____ 12. a type of nucleic acid that includes three major types

_____ 13. a process that occurs at ribosomes where proteins are made from the information found in RNA

Skills Worksheet

Test Prep Pretest

Complete each statement by writing the correct term or phrase in the space provided.

1. In 1928, Griffith found that the ability to cause disease could be transferred between strains of bacteria due to the process of _____.

2. Avery's experiments demonstrated that DNA, and not protein or RNA, is the _____ material.

3. After infecting *Escherichia coli* bacteria with P-labeled phages, Hershey and Chase traced the ^{32}P. The scientists found most of the radioactive substance in the _____.

4. Watson and Crick used the X-ray _____ photographs of Wilkins and Franklin to build their model of DNA.

5. The process of making new DNA is called _____.

6. The Y-shaped area formed when the double helix separates during replication is called a _____ _____.

7. DNA replication occurs before a _____ _____.

8. Eukaryotic DNA contains many replication forks working in concert, whereas prokaryotic DNA contains only _____ replication forks during replication.

9. Proteins that catalyze the formation of a DNA molecule are _____ _____.

10. Gene expression occurs through transcription and _____.

11. _____ places the amino acids on the growing polypeptide chains.

12. Messenger RNA is complementary to the _____ sequence.

| Test Prep Pretest *continued*

In the space provided, write the letter of the description that best matches the term or phrase.

_____13. transformation

_____14. replication

_____15. DNA helicase

_____16. Wilkins and Frank

_____17. Watson and Crick

_____18. RNA polymerase

_____19. tRNA

_____20. RNA

a. discovered the three-dimensional structure of DNA with the help of other scientists

b. binds to a genes promoter

c. developed high quality X-ray diffraction photographs of DNA

d. results in two DNA molecules that are identical to the original DNA molecule

e. results in a change in a cell's genotype

f. contains an anticodon and an amino acid binding site

g. contains uracil instead of thymine

h. unwinds the two DNA strands during replication

Read each question, and write your answer in the space provided.

21. Relate the role of base-pairing rules to the structure of DNA.

_____

_____

_____

_____

22. Describe the components of a nucleotide in DNA.

_____

_____

23. What happened when Griffith mixed harmless living R bacteria with harmless heat-killed S bacteria and then injected mice with this mixture?

_____

_____

_____

Test Prep Pretest *continued*

24. Why did Hershey and Chase use radioactive elements in their experiments?

_____

_____

_____

_____

25. Explain how DNA polymerase "proofreads" a new DNA strand.

_____

_____

_____

_____

_____

26. Describe the role of DNA helicases during replication.

_____

_____

_____

27. Explain how RNA differs from DNA.

_____

_____

_____

_____

28. Describe the functions of RNA.

_____

_____

_____

_____

_____

Questions 29–31 refer to the figure below.

DNA   RNA   Protein

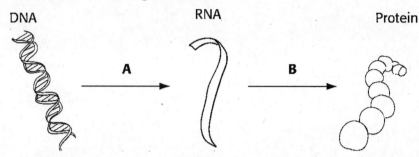

29. The processing of information from DNA into proteins, as shown above, is

    referred to as _____ _____.

30. Stage A is called _____.

31. Stage B is called _____.

In the space provided, write the letter of the term or phrase that best completes each statement or best answers each question.

_____ 32. Which of the following represents the codons that correspond to this
        segment of DNA: TAT—CAG—GAT?
        a. AUA—GUC—CUA   c. AUAGU—CCUA
        b. ATA—GTC—CTA   d. ACA—CUC—GUA

_____ 33. Which of the following are the anticodons that correspond to the
        mRNA codons CAG—ACU—UUU?
        a. GTC—TGA—AAA
        b. GUC—UGA—AAA
        c. glutamine—threonine—phenylalanine
        d. GAC—UCA—AAA

Name _____ Class _____ Date _____

Vocabulary Review

In the space provided, write the letter of the term that best completes each statement.

_____ 1. A unit of adjacent genes including regulatory genes and closely related structural genes is called a(n)
   a. transcription factor.   c. operon.
   b. mutation.   d. promoter.

_____ 2. A genetic structure in bacteria that is separate from the chromosome and can replicate on its own is a(n)
   a. plasmid.   c. exon.
   b. domain.   d. transposon.

_____ 3. All the genetic information in an organism is referred to its
   a. chromosome.   c. chromatid.
   b. genome.   d. plasmid.

_____ 4. A genetic sequence that can randomly move between different genomes is a
   a. chromatid.   c. polyploidy.
   b. homeotic gene.   d. transposon.

_____ 5. A genetically controlled process that leads to cell death is
   a. development.   c. apoptosis.
   b. hox.   d. cell differentiation.

_____ 6. A protein that regulates gene expression is called a(n)
   a. mutation.   c. non coding sequence.
   b. operon.   d. transcription factor.

_____ 7. A non coding segment of DNA is a(n)
   a. intron.   c. nondisjunction.
   b. transposon.   d. cyclin.

_____ 8. A segment of DNA that can be translated is a(n)
   a. transposon.   c. exon.
   b. nondisjunction.   d. intron.

_____ 9. The process by which a cell becomes specialized is cell
   a. differentiation.   c. apoptosis.
   b. insertion.   d. mutation.

_____ 10. A change in the structure or number of genes is a(n)
   a. exon.   c. genome.
   b. mutation.   d. intron.

_____11. The failure of homologous chromosomes to separate during meiosis is
   a. polyploidy.
   b. nondisjunction.
   c. cell differentiation.
   d. genetic switch.

_____12. A distinctive functional unit in a protein is called a(n)
   a. intron.
   b. exon.
   c. domain.
   d. homeobox.

_____13. Having more than one set of chromosomes is called
   a. protein sorting.
   b. polyploidy.
   c. endosymbiosis.
   d. RNA splicing.

Skills Worksheet

Test Prep Pretest

In the space provided, write the letter of the term or phrase that best completes each statement or best answers each question.

_____ 1. Gene regulation in eukaryotes causes
   a. genes to be expressed differently depending on the cellular environment.
   b. genes to be expressed in every kind of environment.
   c. genes to be expressed differently depending on the number of genes that occur together.
   d. None of the above.

_____ 2. What is the name of the proteins that regulate gene expression?
   a. regulators
   b. transcription factors
   c. gene animators
   d. promoters

_____ 3. The part of a protein that is chemically active is called
   a. the active domain.
   b. the subunit.
   c. a regulatory protein.
   d. RNA splicing.

_____ 4. In what kinds of cells do mutations occur that can be transmitted to offspring?
   a. body cells
   b. gametes
   c. reproductive cells
   d. Both (b) and (c)

_____ 5. A mutation that can moves a gene to a new location is called a(n)
   a. point mutation.
   b. insertion.
   c. transposon.
   d. deletion.

_____ 6. Gene mutations that result in cancer often affect
   a. genes that control blood clotting.
   b. genes that affect tRNA and rRNA production.
   c. genes that control cell growth and specialization.
   d. Both (a) and (b)

_____ 7. Which of the following is *not* an example of large-scale genetic change?
   a. nondisjunction
   b. polyploidy
   c. crossover
   d. point mutation

Complete each statement by writing the correct term or phrase in the space provided.

8. When lactose is absent, _____ _____ can

bind to the promoter and transcription can occur in the *lac* operon.

9. The *lac* operon is switched off when a protein called a(n)

_____ is bound to the operator.

10. In eukaryotic gene regulation, proteins called _____

_____ help arrange RNA polymerases in the correct

position on the promoter.

11. In eukaryotes, long segments of nucleotides with no coding information are

called _____.

12. In eukaryotes, the portions of a gene that are actually translated into proteins

are called _____.

13. Insertions, deletions and point mutations are types of

_____ mutations.

14. Duplications, deletions, inversions, and gene rearrangements are types of

_____ mutations.

Test Prep Pretest *continued*

Read each question, and write your answer in the space provided.

15. Explain the difference between a nonsense mutation and a missense mutation.

_____

_____

_____

16. What happens when nondisjunction takes place during cell division?

_____

_____

_____

_____

17. Describe what happens during apoptosis.

_____

_____

_____

18. What is the *lac* operon?

_____

_____

_____

19. Explain why gene regulation in eukaryotic cells is more complex than in prokaryotic cells.

_____

_____

_____

_____

_____

Test Prep Pretest *continued*

20. Describe the three ways that DNA mutation can alter genetic material.

_____

_____

_____

_____

_____

Name _____ Class _____ Date _____

Skills Worksheet

Vocabulary Review

In the space provided, write the letter of the description that best matches each term.

_____ 1. microarray

_____ 2. clone

_____ 3. stem cell

_____ 4. bioinformatics

_____ 5. genomics

_____ 6. restriction enzymes

_____ 7. DNA polymorphisms

_____ 8. genome mapping

_____ 9. electrophoresis

_____ 10. polymerase chain reaction

_____ 11. genetic engineering

_____ 12. DNA fingerprint

_____ 13. DNA sequencing

_____ 14. genetic library

_____ 15. recombinant DNA

a. a technique used to make many copies of a piece of DNA

b. the study of entire genomes

c. a device that can test for the presence or absence of gene activity

d. variations in DNA sequences

e. bacterial enzymes that recognize and bind to specific short sequences of DNA, then cut the DNA at specific sites within the sequences

f. a technique that uses an electrical field within a gel to separate molecules by their size and charge

g. a cell that can continuously divide and differentiate into specialized cell types

h. to make a piece of DNA or organism genetically identical to a preexisting one

i. the application of information technology in biology

j. the process of determining the relative positions of genes in a genome

k. the process of changing the genetic material of an organism

l. the process of determining the order of every nucleotide in a piece of genetic material

m. DNA made from two or more different sources

n. a unique pattern of DNA banding

o. a collection of genetic sequence clones that represent all the genes in a specific genome

Skills Worksheet

Test Prep Pretest

In the space provided, write the letter of the term or phrase that best completes each statement or best answers each question.

_____ 1. How are genetically engineered vaccines different from traditional vaccines?
   a. They are ineffective.
   b. They cause only a mild form of the disease.
   c. They reduce the risk of transmitting the disease to the person injected.
   d. They cause the immune system to make antibodies.

_____ 2. An antibiotic is used in genetic engineering experiments as a way to
   a. identify bacteria that have taken up the recombined plasmid.
   b. produce stronger strains of bacteria.
   c. prevent the cultures from becoming infected with bacteria.
   d. kill cell clones that contain recombinant DNA.

_____ 3. What does a microarray contain?
   a. different gene sequences
   b. harmless viruses
   c. human genes
   d. encoded antibodies

_____ 4. Gene technology is used to improve agriculture and medicine in which of the following ways?
   a. altering muscle percentage in animals
   b. making more nutritious food crops
   c. producing milk containing human proteins by adding human genes to farm-animal genes
   d. All of the above

Questions 5–7 refer to the figure below, which shows the steps of a genetic engineering experiment using DNA from a human insulin gene.

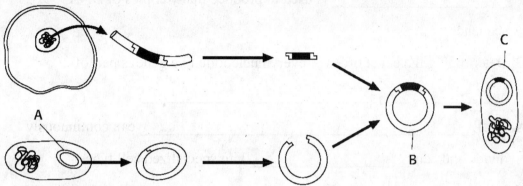

_____ 5. The structure labeled *A* is called
    a. plasmid DNA.
    b. a vector.
    c. a restriction enzyme.
    d. Both (a) and (b)

_____ 6. In *B*, the DNA of the gene and the vector are
    a. cloned.
    b. isolated.
    c. recombined.
    d. the same size.

_____ 7. In *C*, the
    a. gene is cloned.
    b. cells are screened for antibiotic resistance.
    c. recombined plasmid DNA is inserted into the bacterium.
    d. DNA is cut.

Complete each statement by writing the correct term or phrase in the space provided.

8. _____ _____ informs people about

the risk of genetic problems that could affect them or their children.

9. DNA that has recombined by genetic engineering is called

_____ DNA.

10. Any two fragments of DNA cut by the same restriction enzyme can pair

because their ends are _____ .

11. Genetic engineering has benefited humans afflicted with diabetes by

developing bacteria that produce _____ .

12. _____ _____

    _____ is used to produce many copies of DNA

    sequences.

13. Determining the exact order of every nucleotide in a gene is part of

    _____ _____.

14. A(n) _____ _____ can continuously

    divide and can _____ into specialized cell types.

15. In genetic engineering, the enzyme _____

    _____ helps DNA fragments bond together.

Read each question, and write your answer in the space provided.

16. Describe the Human Genome Project.

    _____

    _____

    _____

    _____

17. List two ways in which DNA fingerprints are used.

    _____

    _____

    _____

18. Explain why the development of genetically engineered proteins has been
    important to pharmaceutical companies.

    _____

    _____

    _____

    _____

    _____

    _____

Test Prep Pretest *continued*

19. What is a Southern Blot specifically used for?

_____

_____

20. Identify two problems with cloning animals.

_____

_____

_____

_____

Name _____ Class _____ Date _____

Vocabulary Review

In the space provided, write the letter of the term or phrase that best completes each statement.

_____ 1. The process in which organisms with traits well suited to an environment are more likely to survive and to produce offspring is
a. origin of species.
b. genetic principles.
c. natural selection.

_____ 2. In biology, the formation of species is called
a. speciation.
b. adaptation.
c. artificial selection.

_____ 3. The process by which species change over time is called
a. homologous.
b. evolution.
c. speciation.

_____ 4. The process by which a species becomes better suited to its environment is
a. speciation.
b. adaptation.
c. extinction.

_____ 5. Structures that share a common ancestry are
a. not related.
b. homologous.
c. analogous.

_____ 6. Selection done by humans is
a. natural selection.
b. artificial selection.
c. adaptation.

_____ 7. Remnants of organisms living in the past are
a. gradualism.
b. fossils.
c. adaptations.

Skills Worksheet

Test Prep Pretest

In the space provided, write the letter of the term or phrase that best completes each statement or best answers each question.

_____ 1. On the Galápagos Islands, Darwin saw that the plants and animals closely resembled those found
    a. on islands off the coast of North America.
    b. in South America.
    c. on islands off the coast of Africa.
    d. in South Africa.

_____ 2. Which of the following is a factor in natural selection?
    a. Individuals of a population overproduce.
    b. All populations are genetically diverse.
    c. Individuals better able to adapt to changes leave more offspring.
    d. All of the above

_____ 3. When a population of a species is split in two and the two groups separate for a long period of time, the two groups may become
    a. different families.   c. the same species.
    b. different species.   d. unrelated.

_____ 4. The fossil record provides evidence that
    a. older species from the past gave rise to more-recent species.
    b. all species were formed during Earth's formation and have changed little since then.
    c. the fossilized species have no connection to today's species.
    d. fossils cannot be dated.

_____ 5. Comparing human hemoglobin with the hemoglobin of gorillas, mice, chickens, and frogs reveals that humans have the fewest amino acid differences with
    a. gorillas.   c. chickens.
    b. mice.   d. frogs.

_____ 6. Individuals that are better able to cope with the challenges of their environment tend to
    a. decrease in population over time.
    b. leave more offspring than those more suited to the environment.
    c. leave fewer offspring than those less suited to the environment.
    d. leave more offspring than those less suited to the environment.

Test Prep Pretest *continued*

_____ 7. Which factor does *not* play a role in natural selection?
   a. overproduction
   b. variation
   c. Lamarckian inheritance
   d. adaptation

_____ 8. Different populations of the same species
   a. always become different species over time.
   b. may change enough to become different species.
   c. can no longer interbreed successfully.
   d. will never diverge to become different species.

Questions 9 and 10 refer to the figures below.

Chicken embryo   Human embryo

Pharyngeal pouch   Pharyngeal pouch

Bony tail   Bony tail

_____ 9. Which of the following statements best reflects the evolutionary importance of the figures above?
   a. New genetic instructions have been disregarded in the evolution of vertebrates.
   b. In parts of development, vertebrate embryos show evidence of common ancestry.
   c. The evolutionary history of organisms is seen in transitional embryos.
   d. All adult vertebrates retain tails.

_____10. Which of the following statements is *not* true about anatomy and evolution?
   a. Homologous structures indicate common ancestry of organisms.
   b. The bone patterns making up the forelimbs of tetrapods are similar.
   c. Internal similarities do not indicate shared evolutionary history.
   d. Most vertebrates have four limbs.

| Test Prep Pretest *continued*

Complete each statement by writing the correct term or phrase in the space provided.

11. Evolution is the process by which _____ may change over time.

12. While on the *Beagle,* Darwin read Lyell's book, which contained a detailed account about _____ changes that occur in geological processes on Earth.

13. A trait in a species that results in it being better suited to survive and reproduce in its environment is called a(n)_____.

14. A(n) _____ is a group of individuals that belong to the same species, live in a defined area, and breed with others in the group.

15. The formation of a new species is called _____.

16. Species that shared a common ancestor in the recent past have many

_____ _____ or

_____ sequence similarities.

17. Given that the forelimbs of all vertebrates share the same basic arrangement of bones, forelimbs are said to be _____ structures.

18. Populations evolve, but _____ do not evolve.

19. Some whales have tiny _____ bones as evidence of their land-dwelling mammalian ancestors.

20. Darwin felt that fossils of extinct armadillos that resembled living armadillos were evidence of _____ with

_____.

21. A type of evolution with small scale changes in genes is called

_____.

Test Prep Pretest *continued*

Read each question, and write your answer in the space provided.

22. What was Lamarck's incorrect hypothesis regarding inheritance?

_____

_____

_____

23. Briefly explain the importance of Thomas Malthus's essay on the growth of the human population to Darwin's theory of evolution by natural selection.

_____

_____

_____

_____

24. State three ways Darwin's theory has been updated.

_____

_____

_____

_____

_____

_____

_____

Name _____ Class _____ Date _____

Skills Worksheet

Vocabulary Review

Complete each statement in the space provided by writing the correct term from the list below.

adaptive radiation	genetic drift	polygenic
directional selection	genetic equilibrium	population genetics
divergence	Hardy-Weinberg principle	reproductive isolation
extinction	microevolution	speciation
gene flow	nonrandom mating	stabilizing selection
gene pool	normal distribution	subspecies

1. A trait that is influenced by several genes is called _____.

2. The evolutionary forces include the mutation of genes and

   _____ _____, which is the

   movement of alleles into or out of a population.

3. In small populations, the frequency of an allele can be greatly changed by a

   chance event, such as a fire or landslide. This change in allele frequency is

   called _____ _____.

4. According to the _____-_____

   _____, the frequencies of alleles in a population do not

   change unless evolutionary forces act on the population.

5. If you were to plot the height of everyone in your class on a graph, the values

   would probably form a hill-shaped curve called a(n) _____

   _____.

6. Sometimes, individuals prefer to mate with others that live nearby or are of

   their own phenotype, a situation called _____

   _____.

7. Evolution at the level of genetic change is called _____.

8. When natural selection eliminates extremes at both ends of a range of phenotypes, the frequencies of the intermediate phenotypes increase. This form of selection is called _____

_____.

9. When natural selection causes the frequency of a particular trait to move in one direction, this form of selection is called _____

_____.

10. When a species fails to produce any more descendants,

_____ occurs.

11. A population in which no genetic change is occurring is in a state of

_____ _____.

12. The divergence of multiple lineages into many new species in a specific area and time is called _____ _____.

13. The particular combination of alleles in a population at any one point in time makes up a(n) _____ _____.

14. The study of changes in the numbers and types of alleles in populations is called _____ _____.

15. A state in which two populations can no longer interbreed to produce future offspring is _____ _____.

16. The accumulation of differences between populations is called

_____.

17. A population that differs from, but can interbreed with, other populations of the same species is called a(n) _____.

18. The process of forming new species by evolution from preexisting species is called _____.

Name _____ Class _____ Date _____

Test Prep Pretest

Complete each statement by writing the correct term or phrase in the space provided.

1. A female robin that chooses a mate based on how well he sings is

   demonstrating _____ _____.

2. Migration to or from a population results in _____

   _____.

3. If the graph of the phenotypes of a trait in a population is a hill-shaped curve,

   the trait exhibits a(n) _____ _____.

4. When a recessive allele is present at a frequency of 0.1, only 1 out of 100

   individuals will be homozygous recessive and will display the phenotype

   associated with this allele. However, 18 out of 1,000 individuals will be

   _____ and will carry the allele unexpressed.

5. Over time, change within species leads to the addition of new species while

   some species become _____.

6. The changing of a species that results in its being better suited to its

   environment is called _____.

7. A condition in which two groups of a population have diverged sufficiently

   that they can no longer interbreed is called _____

   _____.

8. Darwin knew about heredity, but he did not know about _____.

9. Microevolution is the study of evolution at the level of _____.

10. One of Charles Darwin's contributions to biology was his careful study of

    _____ _____, such as the many

    colors of a species of flower.

In the space provided, write the letter of the description that best matches each term.

_____11. population genetics

_____12. normal distribution

_____13. phenotypic variation

_____14. genotype

_____15. genetic equilibrium

_____16. reproductive isolation

_____17. subspecies

a. the visible expression of various genotypes

b. a population that has diverged noticeably from other populations

c. a state in which allele frequencies of a population remain the same

d. the study of the changes in numbers and types of alleles in populations

e. a pattern of distribution in which trait values cluster around an average

f. a set of alleles that determine an individual's phenotype

g. a state in which a population is no longer interbreeding with other populations

In the space provided, write the letter of the term or phrase that best completes each statement or best answers each question.

_____ 18. When the individuals of two populations can no longer interbreed, the two populations are considered to be
a. different families.
b. reproductively isolated.
c. the same species.
d. unrelated.

_____ 19. Members of different subspecies
a. are completely different species.
b. have different adaptations than their parent species.
c. can no longer interbreed successfully.
d. will never diverge to become different species.

_____ 20. The Hardy-Weinberg principle
a. can predict genotype frequencies.
b. can predict genetic drift.
c. applies only to large populations.
d. Both (a) and (b)

_____ 21. Natural selection acts directly on which of the following?
a. genotypes
b. phenotypes
c. both phenotypes and genotypes
d. neither phenotypes nor genotypes

_____ 22. In large, randomly mating populations, the frequencies of alleles and
genotypes are likely to remain constant from generation to generation
unless
a. evolutionary forces are absent.
b. evolutionary forces act on the population.
c. the populations are bacterial.
d. the populations are human.

_____ 23. Human height is an example of a
a. single-gene trait.
b. double-gene trait.
c. monogenic trait.
d. polygenic trait.

_____ 24. The range of phenotypes shifts toward one extreme in
a. stabilizing selection.
b. disruptive selection.
c. directional selection.
d. polygenic selection.

_____ 25. Which of the following is not a factor in natural selection?
a. All populations have genetic variation.
b. Individuals of a species cannot compete if they are to survive.
c. All populations depend upon the reproduction of individuals.
d. Individuals tend to produce more offspring than the environment can
support.

Question 26 refers to the figure below.

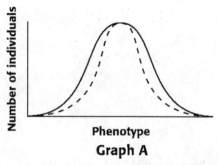

Phenotype
Graph A

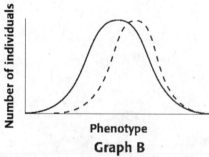

Phenotype
Graph B

26. What type of distribution is represented by the solid-line curve in these
graphs?

_____

_____

Name _____ Class _____ Date _____

Skills Worksheet

Vocabulary Review

Complete each statement by writing the correct term from the list below in the space provided.

analogous character	convergent evolution	order
archaea	derived characters	phylogenetic tree
bacteria	domain	phylogeny
binomial nomenclature	eukaryote	phylum
cladistics	family	prokaryote
cladogram	genus	taxonomy
class	kingdom	

1. The classification level in which classes with similar characteristics are

   grouped is called a(n)_____.

2. Organisms made up of one or more cells with complex internal structure are

   called _____.

3. Reconstructing phylogenies by inferring relationships based on similarities

   derived from a common ancestor without considering the "strength" of a

   character is called _____.

4. The evolutionary history of a species is its _____.

5. Orders with common properties are combined into a(n)

   _____.

6. Similar families are combined into a(n) _____.

7. The classification level in which similar genera are grouped is called a(n)

   _____.

8. A similar feature that evolved through convergent evolution is called a(n)

   _____ _____.

9. In _____ _____, organisms evolve

   similar features independently, often because they live in similar habitats.

10. A(n) _____ is a branching diagram used to show

evolutionary relationships in groups with shared derived characters.

11. The _____ are prokaryotic organisms made of cells that

have a strong exterior cell wall.

12. The most general level of classification is _____.

13. A(n) _____ is a taxonomic category that contains similar

species.

14. Methanogens and extremophiles are examples of _____.

15. Linnaeus developed a system for naming and classifying organisms, which is

called _____.

16. A(n) _____ is an organism that is made up of cells that

lack a nucleus and most other cell organelles.

17. Unique characteristics used in cladistics are called _____

_____.

18. The two-word system for naming organisms is called

_____ _____.

19. A(n) _____ contains many phyla.

20. In phylogenetics, evolutionary relationships are shown in a branching diagram

called a _____ _____.

Skills Worksheet

Test Prep Pretest

In the space provided, write the letter of the term or phrase that best completes each statement or best answers each question.

_____ 1. Although Linnaeus used the Latin polynomial system in his books, he created his own
   a. rules of grammar.
   b. taxonomic categories.
   c. evolutionary systematics.
   d. two-word shorthand system, also in Latin.

_____ 2. Scientists classify organisms by studying their forms and
   a. structures.
   b. size.
   c. method of reproduction.
   d. cladograms.

_____ 3. Cladograms determine evolutionary relationships between organisms by examining
   a. the strength of a character.
   b. the degree of difference between organisms.
   c. shared ancestral characters.
   d. shared derived characters.

_____ 4. All members of the kingdom Animalia are multicellular
   a. autotrophs whose cells have walls.
   b. heterotrophs whose cells have walls.
   c. heterotrophs whose cells lack walls.
   d. autotrophs whose cells lack walls.

_____ 5. Plant cells have cell walls composed of which of the following?
   a. cellulose
   b. chitin
   c. silica
   d. peptidoglycan

_____ 6. The characteristics that scientists use in cladistics are
   a. analogous structures.
   b. shared derived characters.
   c. convergent structures.
   d. shared homologous traits

_____ 7. Scientific names
   a. must have three Latin words and correct Latin grammar.
   b. include the genus and family.
   c. have rules established by British and American biologists.
   d. enable biologists to communicate regardless of their native language.

_____ 8. Bird wings and insect wings are
  a. homologous traits.   c. analogous traits.
  b. derived traits.   d. phylogenetic traits.

_____ 9. Which of the following lists the eight classification levels in proper
  descending order?
  a. domain, kingdom, phylum, class, order, family, genus, species
  b. kingdom, domain, phylum, order, class, family, genus, species
  c. kingdom, phylum, family, class, domain, order, genus, species
  d. phylum, kingdom, domain, class, order, family, genus, species

_____ 10. The scientific naming system requires all of the following *except* that
  a. both words should be underlined or italicized.
  b. the genus is to be capitalized.
  c. the species should be the second word.
  d. the genus is never abbreviated.

**Complete each statement by writing the correct term or phrase in the space
provided.**

11. The only domain that includes multicellular organisms is

  _____.

12. The two kingdoms in which all members are heterotrophs are

  _____ and _____.

13. The naming system developed by Linnaeus is called _____

  _____.

14. One genus can include several _____.

15. Similar features in organisms that do not share a recent common ancestor are

  the result of _____ _____.

16. A(n) _____ is a set of groups that are related by descent

  from a single ancestral lineage.

17. The evolutionary history of a species is called its _____.

Read each question, and write your answer in the space provided.

18. Which classification system would probably be used first if a scientist discovered five unknown plants? Explain.

_____

_____

_____

19. List the six kingdoms, and indicate whether the organisms in each kingdom are prokaryotic or eukaryotic.

_____

_____

_____

_____

_____

Questions 20 and 21 refer to the figure at right, which shows a phylogenetic tree of the six kingdoms.

20. Explain why the kingdom Archaebacteria is located on the branch of the tree that leads to the kingdoms Protista, Animalia, Plantae, and Fungi.

_____

_____

_____

_____

_____

21. Does the phylogenetic tree separate prokaryotes from eukaryotes? Explain.

_____

_____

_____

| Test Prep Pretest *continued*

Questions 22–25 refer to the figure below. The phylogenetic tree shown indicates the evolutionary relationships for a hypothetical group of modern organisms, labeled *1–5*, and their ancestors, labeled *A–E*.

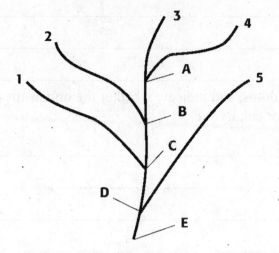

22. Which two modern organisms are likely to be most closely related?

_____

23. What was the most recent common ancestor of the organisms labeled *1* and *5*?

_____

24. Which two modern organisms are likely to be most distantly related?

_____

25. What was the most recent common ancestor of the organisms labeled *1* and *2*?

_____

Skills Worksheet

Vocabulary Review

In the space provided, explain how the terms in each pair differ in meaning.

1. relative dating, radiometric dating

_____

_____

_____

_____

_____

2. fossil record, geologic time scale

_____

_____

_____

_____

_____

In the space provided, write the letter of the description that best matches each term.

_____ 3. cyanobacteria

_____ 4. endosymbiosis

_____ 5. half-life

_____ 6. mass extinction

_____ 7. microspheres

_____ 8. ribozyme

a. used in measuring the rate of decay of a radioactive isotope

b. tiny droplets made of short chains of amino acids in water

c. a type of RNA that can catalyze reactions of organic molecules

d. prokaryotes that carry out photosynthesis

e. the theory that mitochondria and chloroplasts are the descendants of symbiotic bacteria

f. the death of all members of many different species

Skills Worksheet

Test Prep Pretest

In the space provided, write the letter of the term or phrase that best completes each statement or best answers each question.

_____ 1. Scientists estimate that Earth is approximately
   a. 4,000 years old.
   b. 500,000 years old.
   c. 2.5 billion years old.
   d. 4.5 billion years old.

_____ 2. A mechanism for heredity was necessary in order to begin
   a. microspheres.
   b. life.
   c. RNA.
   d. protein.

_____ 3. According to the principle of superposition, which is true of the fossils in the strata of rock in an area like the Grand Canyon?
   a. Deeper strata contain older fossils.
   b. Surface strata contain the oldest fossils.
   c. Older strata contain younger fossils.
   d. All rock strata contain fossils.

_____ 4. Life was able to move from the sea to land because
   a. photosynthesis by cyanobacteria added oxygen to Earth's atmosphere.
   b. ozone was created from the oxygen produced by photosynthesis.
   c. ozone provides a shield from the harsh ultraviolet rays of the sun.
   d. All of the above

_____ 5. The Miller-Urey experiment formed organic molecules when electrical sparks were passed through a mixture of
   a. ammonia and oxygen gases.
   b. methane gas and water vapor.
   c. ammonia and methane gases.
   d. water vapor and oxygen gas.

_____ 6. The Miller-Urey experiment was discarded as a model for the formation of the building blocks of life because the
   a. atmosphere lacked ammonium.
   b. atmosphere lacked ozone.
   c. ocean had not yet formed.
   d. ocean did not yet contain oxygen.

_____ 7. The absolute age of fossils is determined through
   a. mass extinctions.
   b. index fossils.
   c. radiometric dating.
   d. relative dating.

| Test Prep Pretest *continued*

_____ 8. Birds first appeared during the
    a. Precambrian time.   c. Mesozoic Era.
    b. Paleozoic Era.   d. Cenozoic Era.

_____ 9. The great swamps that produced the coal and oil we depend on today
    existed during the
    a. Precambrian time.   c. Mesozoic Era.
    b. Paleozoic Era.   d. Cenozoic Era.

_____ 10. Scientists think the first step toward cellular organization was
    a. nucleotides.   c. microspheres.
    b. colonial algae.   d. RNA enzymes.

In the space provided, write the letter of the description that best matches each term.

_____ 11. prokaryotes

_____ 12. mass extinctions

_____ 13. daughter isotopes

_____ 14. RNA

_____ 15. jawless fishes

_____ 16. flowering plants

_____ 17. multicellularity

_____ 18. half-life

a. the first self-replicating information storage molecule

b. a characteristic of a radioisotope

c. contributed to biodiversity by creating opportunities for new life-forms

d. the first organisms to live on land

e. enabled cell specialization

f. products of radioactive decay

g. the first vertebrates

h. evolved on land during the time that dinosaurs were dominant land animals

Complete each statement by writing the correct term or phrase in the space provided.

19. A meteorite that carried _____ _____

supports the hypothesis that the chemicals for life came to Earth from space.

20. The dominant forms of life on Earth during the Precambrian time were the

_____ .

21. Because of the _____ _____ at the

end of the Permian period, about 96 percent of all species of animals living at

the time became extinct.

22. Fossilized mats of cyanobacteria called _____ are the

most common Precambrian fossils.

23. The first group of animals to live on land was the _____.

Read each question, and write your answer in the space provided.

24. How was the geologic time scale developed?

_____

_____

_____

_____

_____

25. What is the theory of endosymbiosis, and what evidence supports this theory?

_____

_____

_____

_____

_____

_____

Name _____ Class _____ Date _____

Skills Worksheet

Vocabulary Review

Use the terms from the list below to fill in the blanks in the following passage.

bacteriophage   envelope   lytic

capsid   lysogenic   pathogen

The protein coat, or (1) _____, of a virus may contain RNA or DNA,

but not both. Many viruses have a(n) (2) _____, which surrounds the

capsid and helps the virus enter cells. Viruses that infect bacteria are called

(3) _____.

　　　Any agent that causes disease is called a(n) (4) _____.

Viruses cause damage when they reproduce inside cells many times. When the

viruses break out, the cell is destroyed. The cycle of infection, reproduction, and cell

destruction is called the (5) _____ cycle.

　　　During an infection, some viruses stay inside the cells but do not make new

viruses. Instead, the viral genes are inserted into the host chromosome. Whenever the

cell divides, the viral genes also divide, resulting in two infected host cells. This type

of replication cycle is called a(n) (6) _____ cycle.

In the blanks provided, fill in the letters of the term or phrase being described.

7. an extra loop of DNA that contains antibiotic-resistance genes   _ L _ _ _ _ _

8. a poisonous chemical that is produced by bacteria   _ _ _ _ _ N

9. thick-walled structure that helps bacteria survive harsh
   conditions   _ _ D _ _ _ _ _

Vocabulary Review continued

In the space provided, write the letter of the description that best matches the term or phrase.

_____ 10. transduction

_____ 11. peptidoglycan

_____ 12. Gram-positive

_____ 13. resistance

_____ 14. conjugation

_____ 15. transformation

_____ 16. Gram-negative

_____ 17. antibiotic

_____ 18. Koch's postulates

a. bacteria appear pink after staining

b. ability of a bacterium to tolerate antibiotics

c. process in which a virus transfers DNA from one bacterium to another

d. bacteria appear purple after staining

e. a guide for identifying specific pathogens

f. a process in which two organisms exchange genetic material

g. protein-carbohydrate compound found in bacterial cell walls

h. chemical that kills or inhibits the growth of bacteria

i. process in which bacteria take up DNA fragments from their environment

Skills Worksheet

Test Prep Pretest

In the space provided, write the letter of the description that best matches the term or phrase.

_____ 1. capsid

_____ 2. envelope

_____ 3. toxin

_____ 4. bacteriophage

_____ 5. pathogen

_____ 6. lytic cycle

_____ 7. provirus

_____ 8. lysogenic cycle

_____ 9. bacillus

_____ 10. coccus

_____ 11. spirillum

_____ 12. antibiotic

_____ 13. endospore

a. a host chromosome with a viral gene inserted into it

b. causes disease when produced by bacteria

c. a chemical that kills or inhibits the growth of microorganisms

d. a rod-shaped bacterial cell

e. structure that allows bacteria to survive harsh conditions

f. a spiral-shaped bacterial cell

g. a virus's protein coat

h. a cycle in which the viral genome replicates without destroying the host cell

i. a bacterium-infecting virus

j. a cycle of viral infection, replication, and cell destruction

k. a round bacterial cell

l. an agent that causes disease

m. surrounds the capsid of many viruses and helps them enter cells

Complete each statement by writing the correct term or phrase in the space provided.

14. _____ occurs when bacteria take up DNA

fragments from their environment.

15. Viruses must rely on the _____

_____ for reproduction.

16. The capsid of viruses may enclose either the nucleic acid

_____ or the nucleic acid _____.

17. Infectious particles called _____ are

misshapen versions of brain proteins.

18. A(n) _____ is a weakened form of a pathogen that prepares the immune system to recognize and destroy the pathogen.

19. The _____ of *E. coli* have two main functions: to adhere to surfaces and to join bacterial cells prior to

_____.

20. In the presence of hydrogen-rich chemicals, _____ bacteria can manufacture all of their own amino acids and proteins.

Questions 21 and 22 refer to the figure below, which shows the human immunodeficiency virus (HIV).

21. The structure labeled *A* is made of _____ and helps the virus enter the _____ _____.

22. The structure labeled *B* is a(n) _____.

Test Prep Pretest *continued*

Read each question, and write your answer in the space provided.

23. Describe how HIV reproduces.

_____

_____

_____

24. How does *E. coli* reproduce?

_____

_____

_____

25. List Koch's postulates. How do biologists use Koch's postulates?

_____

_____

_____

_____

_____

_____

Name _____ Class _____ Date _____

Vocabulary Review

In the space provided, write the letter of the description that best matches the term or phrase.

_____ 1. gamete

_____ 2. plasmodium

_____ 3. sporophyte generation

_____ 4. algal bloom

_____ 5. pseudopodium

_____ 6. algae

_____ 7. cilia

_____ 8. zygospore

_____ 9. gametophyte generation

_____ 10. binary fission

_____ 11. zygote

_____ 12. flagella

_____ 13. alternation of generations

_____ 14. stipe

_____ 15. red tide

a. short hairlike structures

b. reproductive cycle that includes both meiosis and mitosis

c. haploid phase that produces gametes

d. a diploid zygote with a thick protective wall

e. a mass of cytoplasm that looks like oozing slime

f. haploid reproductive cell

g. diploid phase that produces spores

h. stemlike structure

i. mitosis plus cytokinesis

j. long whiplike structure

k. caused by dinoflagellates

l. a flexible, cytoplasmic extension

m. photosynthetic protists

n. overgrowth of aquatic protists

o. formed by fusion of haploid cells

Skills Worksheet

Test Prep Pretest

In the space provided, write the letter of the description that best matches each term.

_____ 1. green algae

_____ 2. red algae

_____ 3. brown algae

_____ 4. zygospore

_____ 5. cellular slime molds

_____ 6. diatoms

a. a diploid zygote with a thick, protective wall; in *Chlamydomonas* life cycle

b. individual organisms that behave as separate amoebas; gather together to form colonies during times of environmental stress

c. major part of marine plankton; may have given rise to plants

d. multicellular; found in cool marine environments

e. multicellular organisms found in warm ocean waters; their color results from red photosynthetic pigments

f. photosynthetic unicellular protists with silica shells

Complete each statement by writing the correct term in the space provided.

7. Two of the most important features that evolved among the protists are

_____ reproduction and _____.

8. Euglenoids have _____, small organs containing light-sensitive pigments that detect changes in the quality and intensity of light.

9. During conjugation, protists exchange _____.

10. *Ulva* is characterized by two distinct multicellular phases: a diploid, spore-producing phase called the _____ generation and a haploid, gamete-producing phase called the _____ generation.

11. _____ is a form of asexual reproduction in which a part of the parent organism accidentally breaks off and becomes a new organism.

12. _____ are parasitic animal-like protists that cannot move.

Test Prep Pretest *continued*

13. The large brown algae that grow along coasts are known as

_____.

14. The stage of *Plasmodium* that lives in mosquitoes and is injected into humans

is called the _____; the second stage of the *Plasmodium*

life cycle is called the _____.

Questions 15–17 refer to the figure at right, which shows a paramecium.

15. The structures labeled *A* are

_____, which enable

paramecium to move through water.

16. The structure labeled *B* is a(n) _____

_____.

17. The structure labeled *C* is a(n) _____

_____.

**Read each question, and write your answer
in the space provided.**

18. What diseases caused by protists can be
transmitted to humans through drinking water?

_____

_____

19. In what four ways do protists change their environment?

_____

_____

20. Compare the life cycle of *Ulva* with the life cycle of *Chlamydomonas*. What
kinds of protists are *Ulva* and *Chlamydomonas*?

_____

_____

_____

21. List three of the different types of sexual reproduction in protists.

_____

_____

22. What group of protists uses extensions of cytoplasm for locomotion? What are the extensions called?

_____

23. What are diatoms, and how are they beneficial?

_____

_____

_____

_____

_____

_____

24. Describe how diatoms reproduce asexually.

_____

_____

_____

25. How do people become infected with malaria?

_____

_____

_____

Skills Worksheet

Vocabulary Review

In the space provided, write the letter of the description that best matches the term.

_____ 1. chitin

_____ 2. hyphae

_____ 3. mycelium

_____ 4. zygosporangium

_____ 5. saprobe

_____ 6. rhizoid

_____ 7. ascus

_____ 8. dermatophyte

_____ 9. basidium

_____ 10. mycorrhizae

_____ 11. lichen

a. a type of mutualistic relationship formed between fungi and the roots of most plants

b. a thick-walled sexual structure

c. the tough polysaccharide found in the hard outer covering of insects and fungal cell walls

d. a symbiosis between a fungus and a photosynthetic partner

e. the hypha that anchors a fungus to its source of food

f. slender filaments that compose the body of a fungus

g. tangled mass formed by hyphae

h. fungus that absorbs nutrients from dead organisms

i. a saclike structure in which haploid spores are formed

j. a club-shaped sexual reproductive structure

k. fungi that infect the skin and nails

Skills Worksheet

Test Prep Pretest

In the space provided, write the letter of the term or phrase that best completes each statement or best answers each question.

_____ 1. Which of the following is *not* a characteristic of fungi?
   a. filamentous bodies
   b. cell walls made of chitin
   c. chlorophyll
   d. heterotrophic

_____ 2. A mycelium helps a fungus absorb nutrients from its environment because it provides
   a. minerals.
   b. a large surface area.
   c. digestive enzymes.
   d. a small surface area.

Questions 3 and 4 refer to the figure at right.

_____ 3. The fungus shown is a(n)
   a. ascomycete.
   b. basidiomycete.
   c. chytrid.
   d. zygomycete.

_____ 4. The structure labeled *A* in the figure is called a
   a. rhizoid.
   b. spore.
   c. mycelium.
   d. hypha.

_____ 5. The classification of organisms in the four phyla of the kingdom Fungi is based on
   a. food.
   b. digestive structures.
   c. cellular structure.
   d. sexual reproductive structures.

_____ 6. Most fungal spores are formed by
   a. the fusing of hyphae.
   b. the fusing of asci.
   c. mitosis.
   d. None of the above

Complete each statement by underlining the correct term or phrase in the brackets.

7. The tough material found in the cell walls of all fungi is [cellulose / chitin].

8. The slender filaments that make up the bodies of most fungi are called

   [hyphae / mycelium].

Test Prep Pretest *continued*

9. In bread mold, the hyphae that grow into the surface of the bread are called [mycelia / rhizoids].

10. Fungi are important to the environment because they decompose [organic matter / minerals].

11. In lichens, the algal partner provides [minerals / carbohydrates].

Complete each statement by writing the correct term in the space provided.

12. Sexual spores are produced by _____; asexual spores are produced by _____.

13. Fungi secrete digestive _____ that break down organic matter into _____, which are absorbed by the fungus.

14. Sexual reproduction in fungi begins when two _____ of opposite mating types fuse and form a reproductive structure.

15. Fungi form distinctive structures during sexual reproduction. Members of the phylum Zygomycota form _____; members of the phylum Ascomycota form _____; and members of the phylum Basidiomycota form _____.

16. Fungi that absorb nutrients from dead organisms are called _____.

17. Certain fungi play important roles in the nutrition of plants by forming symbiotic associations with their roots, called _____.

18. The underside of a mushroom cap is lined with rows of _____, which contain thousands of club-shaped structures called _____.

19. _____ fungi provide clues about fungal evolution.

20. _____ produces gasohol, a fuel alternative to gasoline.

Read each question, and write your answer in the space provided.

21. Explain why fungal infections are difficult to cure.

_____

_____

_____

22. Distinguish between the formation of sexual spores and asexual spores in fungi. How are the spores alike? How are they different?

_____

_____

_____

_____

_____

23. What does each partner contribute to a mycorrhizal relationship?

_____

_____

_____

24. What does *dikaryotic* mean? How does it relate to fungi?

_____

_____

_____

_____

25. Describe how a mushroom obtains nutrients.

_____

_____

_____

Skills Worksheet

Vocabulary Review

In the space provided, write the letter of the term that best matches each description.

_____ 1. a waxy or fatty and watertight layer that covers a plant's epidermal cells

_____ 2. a structure in seedless plants that produces eggs

_____ 3. a structure in seedless plants that produces sperm

_____ 4. a horizontal underground stem

_____ 5. contains a male gametophyte of a seed plant

_____ 6. the part of the sporophyte in which the female gametophyte develops

_____ 7. the transfer of pollen grains from the male to the female reproductive structure

_____ 8. leaflike structure that is part of a plant embryo

_____ 9. a flower structure that consists of a threadlike filament topped with an anther

_____ 10. a pollen-producing sac

_____ 11. the female reproductive structure of a flower

a. anther

b. antheridium

c. archegonium

d. cotyledon

e. cuticle

f. ovule

g. pistil

h. pollen grain

i. pollination

j. rhizome

k. stamen

In the space provided, explain how the terms in each pair differ in meaning.

12. seed, spore

_____

_____

_____

_____

13. frond, sorus

_____

_____

_____

_____

| Vocabulary Review *continued*

14. fruit, sporangium

_____

_____

_____

_____

15. gymnosperms, angiosperms

_____

_____

_____

_____

16. monocot, dicot

_____

_____

_____

_____

17. gametophyte, sporophyte

_____

_____

_____

_____

Skills Worksheet

Test Prep Pretest

In the space provided, write the letter of the term or phrase that best completes each statement or best answers each question.

_____ 1. The gametophyte of a nonvascular plant produces sperm in a structure called a(n)
    a. sporangium.   c. antheridium.
    b. archegonium.   d. sorus.

_____ 2. In seedless vascular plants, the archegonia and antheridia develop on which of the following?
    a. roots of the gametophytes
    b. tips of the gametophytes
    c. lower surfaces of the sporophytes
    d. lower surfaces of the gametophytes

_____ 3. In gymnosperms, the female cones produce
    a. ovules.   c. ovules and seeds.
    b. pollen.   d. pollen and seeds.

_____ 4. The process by which two sperm fuse with cells of the female gametophyte to produce both a zygote and endosperm is called
    a. meiosis.
    b. double fertilization.
    c. asexual reproduction.
    d. alternation of generations.

_____ 5. The male reproductive parts of a flower are called
    a. petals.   c. sepals.
    b. stamens.   d. pistils.

_____ 6. Growing new potato plants from the tubers of a parent plant is one example of which of the following?
    a. vegetative reproduction   c. sexual reproduction
    b. double fertilization   d. self pollination

_____ 7. All of the following are a means of asexual reproduction in plants *except*
    a. archegonia.   c. stolons.
    b. bulbs.   d. tubers.

_____ 8. The most successful group of plants on land are the
    a. ferns.   c. angiosperms.
    b. mosses.   d. gymnosperms.

In the space provided, write the letter of the term that best matches each description.

_____ 9. reproductive structure of angiosperms

_____ 10. dominant stage in a vascular plant's life cycle

_____ 11. innermost whorl of a flower; produces ovules

_____ 12. pollen-producing sac at the top of a stamen

_____ 13. outermost whorl of a flower

_____ 14. male gametophyte of a seed plant

_____ 15. dominant stage in a nonvascular plant's life cycle

_____ 16. whorl of a flower that attracts insects

a. flower
b. sporophyte
c. pollen grain
d. petals
e. sepals
f. anther
g. pistil
h. gametophyte

Complete each statement by writing the correct term or phrase in the space provided.

17. The transfer of pollen grains from a male reproductive structure of a plant to a female reproductive structure of a plant is _____.

18. The gametophytes of _____ _____ are so small that they are microscopic.

19. Spores are produced in a(n) _____ in mosses. A cluster of these forms a(n) _____ in ferns.

20. Seed plants whose seeds do not develop within a fruit are called

_____.

21. Ferns have mature leaves called _____ and coiled young leaves called _____.

22. A(n) _____ is the part of an angiosperm that contains seeds, and a vegetative part is any nonreproductive part of an angiosperm that can be used to reproduce _____.

23. In angiosperms, a sperm cell fuses with two other haploid cells to form a(n) _____ cell that develops into _____.

Name _____ Class _____ Date _____

Questions 24–27 refer to the figure below, which shows the life cycle of a plant.

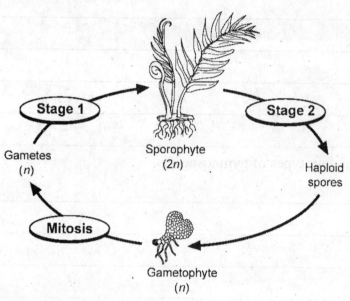

In the space provided, write the letter of the term or phrase that best completes each statement or best answers each question.

_____24. What process occurs at Stage 1?
    a. mitosis   c. fertilization
    b. meiosis   d. cell division

_____25. The structures produced by Stage 1 are
    a. spore capsules.
    b. diploid spores.
    c. haploid spores.
    d. zygotes.

_____26. What process occurs at Stage 2?
    a. fertilization   c. meiosis
    b. pollination   d. mitosis

_____27. The life cycle above is called
    a. a haploid life cycle.
    b. alternation of generations.
    c. a diploid life cycle.
    d. an incomplete life cycle.

Test Prep Pretest *continued*

Read each question, and write your answer in the space provided.

28. List four ways that seeds have influenced the evolution of plants on land.

_____

_____

_____

_____

29. Describe the four types of gymnosperms.

_____

_____

_____

_____

30. Describe the two types of angiosperms, and list two examples of each.

_____

_____

_____

_____

_____

Skills Worksheet

Vocabulary Review

In the space provided, write the letter of the description that best matches each term.

_____ 1. blade

_____ 2. germination

_____ 3. guard cell

_____ 4. heartwood

_____ 5. meristem

_____ 6. mesophyll

_____ 7. petiole

_____ 8. phloem

_____ 9. pith

_____ 10. primary growth

_____ 11. sapwood

_____ 12. secondary growth

_____ 13. stoma

_____ 14. vascular bundle

_____ 15. xylem

a. causes a plant's stems and roots to thicken

b. permits plants to exchange oxygen and carbon dioxide

c. one of a pair of specialized cells that open and close the stomata

d. the broad, flat portion of a typical leaf

e. occurs at the tips of stems and roots

f. the nonconducting older wood in a tree trunk

g. ground tissue found in the center of a dicot stem

h. a strand that contains both xylem and phloem tissue

i. the beginning of plant growth from a seed or spore

j. the part of a tree trunk that is active in transporting water and nutrients

k. tissue made of hard-walled cells that transport water and mineral nutrients

l. ground tissue found in the center of a leaf

m. the stalk that attaches a leaf to a plant's stem

n. an area of undifferentiated plant cells that are capable of dividing

o. tissue made of soft-walled cells that transport organic nutrients

Complete each statement by writing the correct term in the space provided.

16. A type of tissue called _____ tissue forms the protective outer layer of a plant.

17. A type of tissue called _____ tissue makes up much of the inside of the nonwoody parts of a plant.

Vocabulary Review *continued*

18. A type of tissue called _____ tissue functions in the

transport of materials and the support of a plant.

In the space provided, explain how the terms in the pair differ in meaning.

19. apical meristem, lateral meristem

_____

_____

_____

_____

_____

_____

_____

Skills Worksheet

Test Prep Pretest

In the space provided, write the letter of the term or phrase that best completes each statement or best answers each question.

_____ 1. What type of tissue forms the protective outer layers of the plant?
    a. ground   c. dermal
    b. xylem   d. phloem

_____ 2. The primary photosynthetic organs of plants are the
    a. leaves.   c. roots.
    b. stems.   d. flowers.

_____ 3. The leaves, stems, and roots of a plant contain
    a. only one kind of tissue.   c. all three kinds of tissues.
    b. only two kinds of tissue.   d. None of the above

_____ 4. Tendrils are leaves that are specialized for
    a. photosynthesis.   c. climbing.
    b. protection.   d. reproduction.

In the space provided, write the letter of the description that best matches each term.

_____ 5. stomata

_____ 6. dermal tissue

_____ 7. tracheids

_____ 8. petiole

_____ 9. palisade layer

_____ 10. node

_____ 11. pith

_____ 12. vascular bundle

_____ 13. cork

_____ 14. root hairs

_____ 15. xylem

_____ 16. phloem

a. place on a stem where a leaf attaches

b. includes the cork

c. contains both xylem and phloem

d. openings in leaves and stems

e. tissue that transports organic nutrients

f. tissue found in the center of a dicot stem

g. tissue that transports water

h. slender extensions of the epidermis behind the root cap

i. type of cells found in xylem

j. layer of cells filled with waterproof chemical

k. stalk that attaches a leaf to a stem

l. slender cells that are packed with chloroplasts

Test Prep Pretest *continued*

Complete each statement by writing the correct term or phrase in the space provided.

17. When specialized cells called _____ _____ change shape, stomata open and close.

18. The part of a plant's body that grows mostly upward is called the _____; the part that grows downward is called the _____.

19. The cells that carry out metabolic functions for the sieve-tube cells of phloem are called _____ _____.

20. Apical meristems are located at the _____ of _____ and _____.

21. The plant tissues that result from primary growth are known as _____ _____.

22. Dermal tissue prevents water loss, and it also functions in _____ exchange and the absorption of _____ and _____.

23. Ground tissue stores water, _____, and _____, and it contains and supports a plant's _____ tissue.

Read each question, and write your answer in the space provided.

24. What must happen before a seed can germinate?

_____

_____

_____

_____

_____

Test Prep Pretest *continued*

25. Differentiate between nonwoody stems and woody stems.

_____

_____

_____

_____

_____

_____

_____

_____

Name _____ Class _____ Date _____

Vocabulary Review

Complete each statement by writing the correct term in the space provided.

1. The loss of water vapor from a plant is called _____.

2. The movement of organic compounds within a plant from a source to a sink is

   called _____.

3. The chemical substances auxins, ethylene, and gibberellins act as

   _____ in plants.

In the space provided, write the letter of the description that best matches each term.

_____ 4. dormancy

_____ 5. gravitropism

_____ 6. nastic movement

_____ 7. photoperiodism

_____ 8. phototropism

_____ 9. thigmotropism

_____ 10. tropism

a. a response in which a plant grows either toward or away from a stimulus

b. causes a plant to grow toward or away from light

c. a plant growth response to touch

d. a response that does not depend on the direction of a stimulus

e. causes parts of a plant to grow toward or away from the pull of gravity

f. a state of reduced metabolism that causes growth and development to stop

g. a response to seasonal changes in the length of days and nights

Skills Worksheet

Test Prep Pretest

In the space provided, write the letter of the term or phrase that best completes each statement or best answers each question.

_____ 1. The Dutch biologist Frits Went showed that the bending of plants toward light is caused by a chemical called
    a. auxin.   c. nitrogen.
    b. agar.   d. ethylene.

_____ 2. A tropism is a growth response
    a. toward light.   c. toward or away from a stimulus.
    b. to touch.   d. toward gravity.

_____ 3. Many of a plant's responses to environmental stimuli are caused by
    a. the length of the nights.   c. temperature.
    b. hormones.   d. All of the above

_____ 4. Which of the following is *not* a mineral nutrient that plants need?
    a. auxin   c. nitrogen
    b. sulfur   d. magnesium

_____ 5. When stomata are open, water vapor diffuses out of a leaf in a process called
    a. photosynthesis.   c. osmosis.
    b. germination.   d. transpiration.

_____ 6. Water will keep moving upward in a plant as long as there is an unbroken column of water in the
    a. phloem.   c. roots.
    b. xylem.   d. stomata.

Complete each statement by writing the correct term or phrase in the space provided.

7. The closing of a "trap" on a Venus' flytrap whenever a fly lands anywhere on

one of them is a(n) _____ _____.

8. The loss of water vapor by _____ creates a pull that

draws water up through the stem and into the leaves.

9. Roots take in water from the soil by the process called

_____.

Test Prep Pretest *continued*

10. A gaseous compound that _____ fruit ripening and

loosens the fruit of cherries, blackberries, and blueberries is

_____.

11. A condition in which a seed or a plant remains inactive even when conditions

are suitable for growth is called _____.

In the space provided, write the letter of the description that best matches each term.

_____ 12. abscisic acid

_____ 13. auxin

_____ 14. calcium

_____ 15. cytokinin

_____ 16. gibberellin

_____ 17. magnesium

_____ 18. nitrogen

_____ 19. phosphorus

_____ 20. potassium

a. mineral nutrient used to make chlorophyll

b. hormone that stimulates cell division

c. mineral nutrient used to make proteins

d. hormone that slows growth in plants

e. mineral nutrient used in active transport

f. hormone that causes plant stems to bend

g. mineral nutrient used to make cell walls

h. hormone that causes fruit development and seed germination

i. mineral nutrient also used to make ATP

Read each question, and write your answer in the space provided.

21. Summarize how Frits Went demonstrated the presence of the chemical auxin in a shoot tip.

_____

_____

_____

_____

_____

_____

_____

22. Compare and contrast transpiration and translocation in plants.

_____

_____

_____

23. Trace the movement of water through a plant.

_____

_____

_____

_____

Questions 24 and 25 refer to the figure below, which shows a growing plant.

24. Which letters indicate areas of the plant in which its growth was affected by a negative gravitropism? Why is this response important?

_____

_____

25. Which letter indicates a part of the plant that developed as it did because of a positive phototropism? Explain.

_____

_____

Skills Worksheet

Vocabulary Review

Write the correct term from the list below in the space next to its definition.

amniotic egg	coelom	gastrovascular cavity
blastula	deuterostome	gastrulation
cephalization	endoskeleton	heterotroph
cleavage	exoskeleton	hydrostatic skeleton

_____ 1. the series of cell divisions that occur immediately after an egg is fertilized

_____ 2. an animal that obtains food by eating other organisms or their byproducts

_____ 3. a cavity that is filled with water and that has a support function

_____ 4. a body cavity that contains the internal organs

_____ 5. a hard, external, supporting structure that develops from the ectoderm

_____ 6. the formation of embryonic germ layers

_____ 7. a cavity that serves both digestive and circulatory purposes in some animals

_____ 8. a type of egg that is produced by reptiles, birds, and some mammals that contains a large amount of yolk.

_____ 9. an animal whose mouth does *not* derive from the blastopore

_____ 10. the stage of an embryo before gastrulation

_____ 11. the concentration of nerve tissue and sensory organs at the anterior end of an organism

_____ 12. an internal skeleton made of bone and cartilage

| Vocabulary Review *continued*

In the space provided, write the letter of the description that best matches each term.

_____ 13. invertebrate

_____ 14. notochord

_____ 15. protostome

_____ 16. therapsid

_____ 17. vertebrate

a. the embryonic blastopore develops into the mouth

b. animal that does not have a backbone

c. an animal that has a backbone

d. a stiff rod that develops along the back of an embryo

e. a member of the extinct order of mammal-like reptiles that likely gave rise to mammals

Complete each statement by writing the correct term or phrase in the space provided.

18. An animal is a multicellular _____ that has cells that lack cell walls.

19. A bony skull and an internal skeleton made of bone or cartilage are two characteristics shared by _____.

20. Most bilaterally symmetrical animals have an anterior concentration of sensory structures called _____.

21. The end result of cleavage is a(n) _____.

22. During _____, the blastula begins to collapse inward.

23. An extinct order of mammal-like reptiles that were dominant on land briefly and that gave rise to mammals are called _____.

24. A rod-shaped supporting axis found in the dorsal part of the embryos of all chordates, including vertebrates, is the _____.

25. The major evolutionary innovations that first appeared in reptiles include watertight, scale-covered skin and the _____

_____.

Skills Worksheet

Test Prep Pretest

In the space provided, write the letter of the term or phrase that best completes each statement.

_____ 1. The cells of all animals are organized into structural and functional units called tissues *except* for the cells of
    a. sponges.
    b. cnidarians.
    c. flatworms.
    d. roundworms.

_____ 2. Animals are multicellular, heterotrophic organisms with cells that lack
    a. mitochondria.
    b. cell membranes.
    c. cell walls.
    d. ribosomes.

_____ 3. An animal in which the space between the body wall and gut is completely filled with tissues and organs is called a(n)
    a. acoelomate.
    b. pseudocoelomate.
    c. coelomate.
    d. vertebrate.

_____ 4. An animal whose gut has only one opening has a(n)
    a. intervascular cavity.
    b. gastrovascular cavity.
    c. specialized digestive tract.
    d. one-way digestive system.

_____ 5. Animals that do not move, but catch food as it drifts by in the water are
    a. predators.
    b. scavengers.
    c. detritus feeders.
    d. filter feeders.

_____ 6. The vast majority of animals are
    a. mammals.
    b. invertebrates.
    c. vertebrates.
    d. aquatic.

_____ 7. The largest structure on Earth that was built by living organisms is the
    a. Great Wall of China.
    b. Great Barrier Reef.
    c. Grand Canyon.
    d. Grand Coulee Dam.

_____ 8. The backbone supports and protects the
    a. heart and lungs.
    b. liver and kidneys.
    c. abdominal muscles.
    d. dorsal nerve cord.

Complete each statement by writing the correct term or phrase in the space provided.

9. Without a(n) _____ _____, an

   animal could not eliminate the waste products of cellular metabolism.

10. The _____ _____ of an earthworm

   is formed from a fluid contained under pressure in a closed cavity.

11. A(n) _____ develops after a zygote undergoes cell

   division to form a hollow ball of cells.

12. A bilaterally symmetrical animal can be one of three basic kinds of internal

   body plans, coelomate, acoelomate, and _____.

13. Muscles, most of the skeleton, the circulatory system, reproductive organs, and

   excretory organs arise from the primary tissue layer called

   _____.

14. A sea anemone's body plan is an example of _____

   _____ because its body parts are arranged around a

   central axis.

15. Except for mollusks, coelomate animals are composed of a series of repeating,

   similar units called _____.

16. The ectoderm, endoderm, and mesoderm are called

   _____ _____

   _____ because they give rise to all the tissues and organs

   of an adult body.

17. Jointed appendages allow animals to perform complex

   _____, such as defensive displays.

18. The _____ _____ of aquatic

   chordate embryos develop into gill structures.

19. In most vertebrates, the stiff rod called the _____ is

present only in the embryo.

20. Today, almost all large land animals are _____.

Read each question, and write your answer in the space provided.

21. Describe three ways in which the lives of humans are connected to other
    animals.

    _____

    _____

    _____

22. Describe the advantage of an excretory system for terrestrial animals.

    _____

    _____

    _____

23. Compare and contrast the complex nervous system of a vertebrate with the
    nerve net of a simple invertebrate, such as a jellyfish.

    _____

    _____

    _____

24. Discuss the major evolutionary change of bilateral symmetry. How did this
    body plan support the complex development of animals?

    _____

    _____

    _____

25. Explain how the first amphibians were so successful out of the water when
    their limbs were not very efficient for moving on land.

    _____

    _____

    _____

Skills Worksheet

Vocabulary Review

In the space provided, write the letter of the description that best matches each term.

_____ 1. pseudocoelom

_____ 2. choanocytes

_____ 3. amoebocytes

_____ 4. spongin

_____ 5. spicules

_____ 6. medusa

_____ 7. polyp

_____ 8. cnidocytes

_____ 9. nematocyst

_____ 10. planula

_____ 11. proglottids

a. sponge cells that have irregular amoebalike shapes

b. resilient flexible protein fiber

c. free-floating life form of a cnidarian

d. stinging cells located on tentacles of cnidarians

e. body cavity that forms between the mesoderm and endoderm in roundworms

f. body form of a cnidarian that is attached to a rock or some other object

g. used to inject a toxin into prey

h. flagellated cells also known as collar cells

i. free-swimming ciliated larva of a cnidarian

j. body sections of a tapeworm

k. tiny needles of silica or calcium carbonate that form a sponge's skeleton

Complete each statement by underlining the correct term in the brackets.

12. Lining the inside of a sponge is a layer of cells called [amoebocytes / choanocytes].

13. [Amoebocytes / Choanocytes] are cells that move around the body wall of the sponge.

14. The skeleton of most sponges is made of [ostia / spicules].

15. A few sponges have skeletons made of a fibrous protein called [spicules / spongin].

16. The cnidarian body form that is tubelike is the [medusa / polyp].

17. Within each cnidocyte is a threadlike organelle called a(n) [amoebocyte / nematocyst].

18. Tapeworms grow by producing body sections called [flukes / proglottids] that contain reproductive units.

Complete each statement by writing the correct term in the space provided.

19. Calcareous sponges have a hard skeleton made of individual calcium carbonate _____.

20. In the cnidarian *Obelia,* asexual reproduction occurs only in the _____ stage.

21. The _____ of a roundworm serves as a primitive circulatory system.

Skills Worksheet

Test Prep Pretest

In the space provided, write the letter of the term or phrase that best completes each statement or best answers each question.

_____ 1. Support for most sponges is provided by a simple skeleton composed of protein fibers called
a. spicules.
b. spongin.
c. oscula.
d. silica.

_____ 2. Which of the following is characteristic of the roundworm *Ascaris*?
a. The eggs can live in soil for years.
b. The eggs can block ducts leading from organs in the human body, such as the gallbladder.
c. The eggs can travel to the lungs and cause respiratory distress.
d. All of the above

_____ 3. The simplest animal that has a one-way digestive system is the
a. fluke.
b. flatworm.
c. roundworm.
d. cnidarian.

Questions 4–6 refer to the figure below, which shows *Dugesia*.

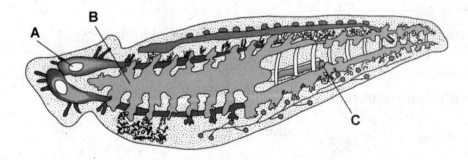

_____ 4. The structure labeled *A* is
a. the brain.
b. a nerve cord.
c. the mouth.
d. an eyespot.

_____ 5. The structure labeled *B* is
a. the gastrovascular cavity.
b. a ciliated cell.
c. a proglottid.
d. a light-sensitive eyespot.

_____ 6. The structure labeled *C* is
a. the intestine.
b. a flame cell.
c. the mouth.
d. the anus.

In the space provided, write the letter of the description that best matches each term.

_____ 7. amoebocyte

_____ 8. osculum

_____ 9. planulae

_____ 10. flame cells

_____ 11. tegument

a. in planarians, specialized cells with beating tufts of cilia that draw water through pores to the outside of the worm's body

b. free-swimming cnidarian larvae

c. an amoebalike cell in a sponge that moves through the body cells, supplying nutrients and removing wastes

d. thick protective covering of cells that protects endoparasites from being digested by their host

e. large opening to the body cavity of a sponge where water is moved out

Complete each statement by writing the correct term in the space provided.

12. Cnidarians have stinging cells called _____ for capturing

prey.

13. The Portuguese man-of-war belongs to a group of cnidarians called

_____.

14. Anthozoans typically have a stalklike body topped by a crown of

_____.

15. *Schistosoma,* sometimes called a blood fluke, must live in a(n)

_____ before it can infect humans.

Read each question, and write your answer in the space provided.

16. Describe how sponge cells get nutrients.

_____

_____

_____

_____

_____

_____

17. Describe sexual reproduction in sponges.

_____

_____

_____

_____

_____

_____

18. Summarize the cellular organization and body forms of *Physalia,* or Portuguese man-of-war.

_____

_____

_____

_____

_____

19. Describe the life cycle of *Obelia*.

_____

_____

_____

_____

_____

_____

20. Describe one simple way that an infection of hookworm can be avoided.

_____

_____

Skills Worksheet

Vocabulary Review

Use the terms from the list below to fill in the blanks in the following passage.

cerebral ganglia	radula	siphons
foot	septa	trochophore
mantle	setae	visceral mass

Mollusks and annelids were probably the first major groups of organisms to develop a true coelom. Another feature shared by many mollusks and annelids is a larval stage called a (1) _____, which develops from the fertilized egg.

Mollusks have many organ systems, which are contained in the (2) _____. A (3) _____ wraps around the visceral mass. Every mollusk has a muscular region called a (4) _____. Many mollusks have one or two shells, which protect their soft bodies. All mollusks, except bivalves, have a tonguelike organ called a (5) _____.

Most bivalves are filter feeders, and many use their muscular foot to dig down into the sand. The cilia on the gills of a bivalve draw in seawater through hollow tubes called (6) _____.

Annelids are easily recognized by their segments, which are visible externally as a series of ringlike structures along the length of their body. Well-developed (7) _____, or primitive brain, are located in one anterior segment. Internal body walls, called (8) _____, separate the segments of most annelids. Most annelids have external bristles called (9) _____.

Skills Worksheet

Test Prep Pretest

In the space provided, write the letter of the term or phrase that best completes each statement or best answers each question.

_____ 1. The fertilized eggs of both mollusks and annelids develop into a distinct larval form called a
  a. polyp.
  b. radula.
  c. trochophore.
  d. nudibranch.

_____ 2. Which of the following is *not* a characteristic of mollusks?
  a. acoelomate body structure
  b. bilateral symmetry
  c. organ systems
  d. three-part body plan

_____ 3. Annelids were the first organisms to exhibit
  a. a true coelom.
  b. organ systems.
  c. bilateral symmetry.
  d. segmentation.

_____ 4. All annelids have a(n)
  a. closed circulatory system and a radula.
  b. closed circulatory system and a mantle.
  c. closed circulatory system and a nerve cord.
  d. open circulatory system and a series of hearts.

_____ 5. Which of the following is *not* a characteristic of annelids?
  a. gills or lungs
  b. organ systems
  c. a highly specialized gut
  d. segmented body

_____ 6. When soil in the digestive tract of an earthworm leaves the crop, it passes to the
  a. pharynx.
  b. gizzard.
  c. esophagus.
  d. anus.

_____ 7. The movement of earthworms requires
  a. muscles lining the interior body wall.
  b. muscle contractions.
  c. traction provided by setae.
  d. All of the above

| Test Prep Pretest *continued*

Complete each statement by writing the correct term in the space provided.

8. When the _____ muscles of a bivalve contract, they cause

the valves to close forcefully.

9. Bivalves feed by sucking seawater through hollow tubes called

_____.

10. The only living cephalopod species that has an outer shell is the

_____.

Questions 11–13 refer to the figure below, which shows the structure of an earthworm.

11. The structure labeled *A,* called the _____, grinds up soil that

the earthworm ingests.

12. The _____ _____, labeled *B,*

coordinates the motor activity of each body segment.

13. The earthworm anchors several of its segments by sinking its

_____, labeled *C,* into the ground.

Test Prep Pretest *continued*

Read each question, and write your answer in the space provided.

14. Why are terrestrial snails less active when the air around them is dry?

_____

_____

_____

15. Describe the function of septa in annelids.

_____

_____

16. How are cephalopods adapted as predators?

_____

_____

_____

_____

17. In what basic way do the annelid and mollusk body plans differ?

_____

_____

18. How are annelids classified?

_____

_____

_____

19. Why do earthworms require a moist environment?

_____

_____

_____

_____

Name _____ Class _____ Date _____

Vocabulary Review

Write the correct term from the list below in the space next to its definition.

appendages	mandible	thorax
cephalothorax	pedipalps	tracheae
chelicerae	spinneret	
Malpighian tubules	spiracle	

_____ 1. structures that extend from the body wall

_____ 2. the midbody region

_____ 3. head fused with thorax

_____ 4. network of tubes through which many arthropods respire

_____ 5. structure through which air enters a terrestrial arthropod's body

_____ 6. excretory units of terrestrial arthropods

_____ 7. mouthparts in the subphylum Chelicerata

_____ 8. pairs of appendages modified to handle prey

_____ 9. appendage that secretes strands of silk

_____ 10. chewing mouthpart in the subphylum Hexapoda

| Vocabulary Review *continued*

In the space provided, write the letter of the description that best matches each term.

_____11. molting

_____12. metamorphosis

_____13. chrysalis

_____14. pupa

_____15. caste

_____16. compound eye

_____17. ossicle

_____18. water-vascular system

_____19. tube foot

_____20. skin gill

a. the role played by an individual in a colony

b. the physical change of a young insect into an adult

c. system of interconnected canals

d. fingerlike projections that create a large surface area for gas exchange

e. stage in complete metamorphosis during which a young insect becomes an adult

f. a protective capsule

g. periodic shedding of exoskeleton

h. calcium-rich plate

i. made of thousands of individual units

j. tiny structure that is used for movement and gripping surfaces

Name _____ Class _____ Date _____

Skills Worksheet

Test Prep Pretest

In the space provided, write the letter of the term or phrase that best completes each statement or best answers each question.

_____ 1. Subphylum Hexapoda includes
  a. insects.
  b. millipedes.
  c. centipedes.
  d. All of the above

_____ 2. All echinoderms have
  a. spine-bearing ossicles.
  b. the ability to move.
  c. a water-vascular system.
  d. a simple brain.

_____ 3. Which of the following characteristics is *not* shared by all insects?
  a. three body sections
  b. wings
  c. three pairs of legs
  d. antennae

_____ 4. The head, thorax, and abdomen of mites
  a. are separate segmented sections.
  b. form two sections, the cephalothorax and the abdomen.
  c. are fused to form a single body.
  d. form two sections, the head and a fused thorax and abdomen.

_____ 5. Spiders produce silk from
  a. spinnerets.
  b. mandibles.
  c. chelicerae.
  d. pedipalps.

In the space provided, write the letter of the description that best matches each term.

_____ 6. exoskeleton

_____ 7. spiracle

_____ 8. Malpighian tubules

_____ 9. madreporite

  a. the shell-like structure that encases the bodies of arthropods
  b. an opening through which water enters an echinoderm's body
  c. fingerlike excretory organs
  d. an opening that functions during respiration in many arthropods

Complete each statement by writing the correct term or phrase in the space provided.

10. An arthropod must shed its _____ to grow.

11. Infected deer ticks may spread _____

   _____ .

12. A larval echinoderm has _____ symmetry.

Name _____ Class _____ Date _____

Questions 13–15 refer to the figures below.

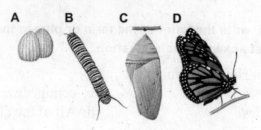

13. The process taking place in the figures above is _____
_____.

14. The stage labeled *D* shows the _____, while the stage
labeled *A* shows the _____.

15. During this process, the _____ is enclosed within a
protective capsule called a(n) _____, labeled *C*.

Read each question, and write your answer in the space provided.

16. Describe the factors that contribute to the evolutionary success of arthropods.

_____

_____

17. List three important characteristics of crustaceans.

_____

_____

18. Distinguish between centipedes and millipedes.

_____

_____

19. Describe the endoskeleton of an echinoderm.

_____

_____

20. List three ways in which the mouthparts of insects are adapted for different
functions.

_____

_____

Name _____ Class _____ Date _____

Skills Worksheet

Vocabulary Review

Complete each statement by writing the correct term or phrase in the space provided.

1. The _____ _____ enables bony fishes to regulate their buoyancy.

2. The _____ _____ allows a fish to perceive its position and rate of movement.

3. The _____ is an efficient respiratory organ due to countercurrent flow.

4. The _____ _____ is an opening at the rear of a fish's cheek cavity.

5. A(n) _____ is an organ that removes metabolic wastes from blood.

6. Movements of the _____ draw water over a fish's gills.

7. A(n) _____ is a fish with a completely symmetrical tail, highly mobile fins, and very thin scales.

8. Along with a small bone, the _____ _____ transmits sound to the inner ear.

9. A(n) _____ is a respiratory organ that allows an animal to get oxygen from the air.

10. The _____ separates the atrium into right and left halves.

11. Oxygen-rich blood is carried from an amphibian's lungs to its heart by the

_____ _____.

12. A larval frog is called a(n) _____.

Skills Worksheet

Test Prep Pretest

In the space provided, write the letter of the term or phrase that best completes each statement.

_____ 1. The major respiratory organ of a fish is the
   a. swim bladder.   c. gill.
   b. lung.   d. operculum.

_____ 2. Depending on the species, fish can reproduce through
   a. internal fertilization.   c. conjugation.
   b. spawning.   d. Both (a) and (b)

_____ 3. Lampreys and hagfishes are the only remaining
   a. jawless fishes.   c. lobe-finned fishes.
   b. cartilaginous fishes.   d. bony fishes.

_____ 4. Caecilians do *not*
   a. use cutaneous respiration.   c. have legs.
   b. lay eggs.   d. bear live young.

_____ 5. Compared with that of a fish, a frog's heart has
   a. fewer chambers.   c. countercurrent flow.
   b. more chambers.   d. Both (b) and (c)

Questions 6–8 refer to the figure at right, which shows the structure of a bony fish.

_____ 6. The structure labeled *A* is the
   a. dorsal fin.
   b. gill filament.
   c. operculum.
   d. lateral line.

_____ 7. The structure labeled *B* is the
   a. pectoral fin.   c. pelvic fin.
   b. operculum.   d. jaw.

_____ 8. The structure labeled *B* helps the fish to
   a. maintain buoyancy.
   b. detect water currents.
   c. move water over its gills.
   d. maintain salt and water balance.

Test Prep Pretest *continued*

Complete each statement by writing the correct term or phrase in the space provided.

9. Fishes breathe by means of _____.

10. A shark's teeth are actually modified _____.

11. In an amphibian, the _____ _____

carry oxygen-rich blood from the lungs to the heart.

12. All fishes have a(n) _____ of either bone or cartilage.

13. In amphibians, one circulatory loop carries blood from the heart to the

_____, while a second loop carries blood to the rest of the body.

14. Hagfishes and lampreys have skeletons of _____.

15. Two structures that maintain a fish's salt and water balance are

_____ and _____.

16. Caecilians detect prey using a(n) _____.

17. The pattern of movement of water and blood through a fish's gills is called

_____ _____.

Read each question, and write your answer in the space provided.

18. Describe countercurrent flow, and tell why it is important to a fish.

_____

_____

_____

_____

19. Explain how a leopard frog is able to hear sounds.

_____

_____

_____

Test Prep Pretest *continued*

20. How do lampreys and hagfishes feed?

_____

_____

_____

21. What information does a bony fish get from its lateral line system?

_____

_____

_____

22. What blood vessels do amphibians have that fish lack? Explain how these vessels provide an advantage to the amphibian.

_____

_____

_____

23. Compare how fertilization takes place in salamanders and caecilians.

_____

_____

_____

24. How do frogs depend on water to complete their life cycles?

_____

_____

_____

Name _____ Class _____ Date _____

Skills Worksheet

Vocabulary Review

In the space provided, write the letter of the description that best matches each term.

_____ 1. ectothermic

_____ 2. endothermic

_____ 3. oviparous

_____ 4. ovoviviparous

_____ 5. carapace

_____ 6. plastron

a. top part of the shell of a turtle

b. body temperature is determined by environmental temperature

c. bottom part of the shell of a turtle

d. condition in which the female retains eggs within her body

e. condition in which the young hatch from eggs outside the mother's body

f. body temperature is maintained by heat generated through metabolism

Complete each statement by writing the correct term or phrase in the space provided.

7. The body of a young bird is covered by _____

_____ that keep it warm.

8. A(n) _____ _____ has branches called

barbs that give it a smooth surface.

9. Snakes have a(n) _____ _____ for

detecting odors.

10. Snakes that suffocate their prey are called _____.

11. Many birds of prey use their _____ to grasp prey.

Test Prep Pretest

In the space provided, write the letter of the term or phrase that best completes each statement or best answers each question.

_____ 1. Tuataras most closely resemble
   a. turtles.
   b. crocodiles.
   c. lizards.
   d. snakes.

_____ 2. In the raising of their young, crocodiles most closely resemble
   a. turtles.
   b. lizards.
   c. snakes.
   d. birds.

_____ 3. All reptiles have the following *except*
   a. a completely divided ventricle.
   b. lungs.
   c. scales.
   d. watertight skin.

_____ 4. Which of the following is *not* true of a turtle's shell?
   a. Vertebrae are fused to the inside of the carapace.
   b. The shell provides support for muscle attachment.
   c. The carapace is always dome shaped.
   d. The shell is made of fused plates of bone.

_____ 5. The second chamber in the stomach of a bald eagle is known as the
   a. crop.
   b. gizzard.
   c. esophagus.
   d. cloaca.

_____ 6. Which of the following is *not* an adaptation for flight?
   a. feathers.
   b. hollow bones.
   c. a keeled breastbone.
   d. two legs.

_____ 7. The most common group of birds are the
   a. birds of prey.
   b. wading birds.
   c. perching birds.
   d. diving birds.

Test Prep Pretest *continued*

In the space provided, write the letter of the description that best matches each term.

_____ 8. crop

_____ 9. plastron

_____ 10. down feather

_____ 11. carapace

_____ 12. gizzard

_____ 13. contour feather

a. grinds food

b. give an adult bird its aerodynamic shape

c. the bottom part of a turtle or tortoise shell

d. the top part of a turtle or tortoise shell

e. stores food

f. conserves body heat

Complete each statement by writing the correct term or phrase in the space provided.

14. Because the _____ in a bird's heart is completely divided, oxygen-rich and oxygen-poor blood are kept completely

_____.

15. An adaptation that indicates a common ancestor for lizards and snakes is a

flexible _____.

16. A long, flattened, rounded bill, as found in _____, is an adaptation for eating grass and other plants.

17. Flight feathers are specialized _____

_____.

18. Most reptiles cannot live in very cold regions because they are

_____.

19. Reptiles, birds, and three species of mammals reproduce by means of

_____ eggs.

20. Birds and many reptiles are _____, meaning their young hatch from eggs laid outside the mother's body.

21. Some species of snakes and lizards are _____, which means the female retains the eggs within her body.

| Test Prep Pretest *continued*

Read each question, and write your answer in the space provided.

22. Describe the skeletal adaptations in reptiles that help them move fast.

_____

_____

_____

_____

23. Describe the shell of a turtle.

_____

_____

_____

_____

24. Which is more efficient, a bird lung or a reptile lung? Explain your answer.

_____

_____

_____

_____

25. Describe the skeletal adaptations for bird flight.

_____

_____

_____

_____

Skills Worksheet

Vocabulary Review

Complete each statement by writing the correct term or phrase in the space provided.

1. An egg-laying mammal is called a(n) _____.

2. The ability to use reflected sound waves to navigate or find objects is called

   _____.

3. Primates that walk upright on two legs are _____.

4. A structure called the _____ allows for the fetus to

   receive nutrients and oxygen from the mother and to get rid of wastes.

5. Mammals have _____ _____ that

   produce milk for nourishing young after their birth.

6. Mammals that have grasping hands and binocular vision are

   _____.

7. The period of time between fertilization and birth is called the

   _____ _____.

Skills Worksheet

Test Prep Pretest

In the space provided, write the letter of the term or phrase that best completes each statement or best answers each question.

_____ 1. Which of the following is *not* a characteristic of mammals?
   a. hair
   b. specialized teeth
   c. ectothermic temperature control
   d. mammary glands

_____ 2. Grizzly bears are able to eat vegetation because they have
   a. a high metabolic rate.
   b. a layer of fat.
   c. a multichambered stomach.
   d. rounded molar teeth with a wrinkled surface.

_____ 3. All female mammals have
   a. a uterus.
   b. mammary glands.
   c. pouches.
   d. nipples.

_____ 4. Incisors are used for
   a. biting and cutting.
   b. stabbing and holding.
   c. crushing and grinding.
   d. Both (a) and (b)

_____ 5. The earliest primates were set apart from their ancestors by having
   a. opposable thumbs and color vision.
   b. large eyes and clawed, unbendable toes.
   c. grasping hands and binocular vision.
   d. color vision and grasping hands.

_____ 6. Compared with modern humans, Neanderthals had
   a. a slightly larger brain, on average.
   b. a much smaller brain, on average.
   c. a taller body.
   d. less prominent brow ridges.

_____ 7. *Homo erectus*
   a. evolved in Africa and remained there.
   b. lived more than 4 million years ago.
   c. gave rise to Homo habilis.
   d. walked upright and used tools.

Complete each statement by writing the correct term or phrase in the space provided.

8. Four types of mammalian teeth are _____,

_____, _____, and

_____.

9. Respiration in mammals is aided by the _____, a sheet of

muscle at the bottom of the rib cage.

10. The length of time between fertilization and birth is the

_____ _____.

11. The duckbill platypus and the echidna are the only living

_____.

12. The most diverse group of mammals in Australia are

_____.

13. *Homo* _____ was associated with tool use and had a(n)

_____ brain than the australopithecines.

14. The size of the _____ brain suggests that a larger brain

was not required for _____ to evolve.

15. The ability of primates to walk _____ probably evolved

in response to environmental changes _____

_____ years ago.

16. Most scientists think that *Homo* _____ or *Homo*

_____ was the direct ancestor of *Homo sapiens*.

17. Modern humans evolved in _____ about

_____ years ago and reached North America from

_____ as early as 15,000 years ago.

Test Prep Pretest *continued*

In the space provided, write the letter of the description that matches the order.

_____18. Order Insectivora

_____19. Order Chiroptera

_____20. Order Proboscidea

_____21. Order Xenarthra

_____22. Order Sirenia

_____23. Order Perissodactyla

_____24. Order Rodentia

a. elongated nose; largest land animals alive today

b. toothless or poorly developed teeth; includes armadillos

c. odd number of toes within their hooves; ungulates

d. only mammals capable of true flight

e. small insect-eaters; may be adapted to burrowing

f. aquatic; related to elephants

g. forty percent of placental mammals

Read each question, and write your answer in the space provided.

25. Compare the degree of development and feeding habits of newborn monotremes, marsupials, and placental mammals.

_____

_____

_____

_____

_____

_____

_____

26. Why are most of the world's marsupials in the Australian region?

_____

_____

_____

Test Prep Pretest *continued*

27. List at least three functions of hair.

_____

_____

_____

28. What general trends are evident in the evolution of hominids? Consider changes to the brain, arms, legs, jaw, and overall height.

_____

_____

_____

_____

_____

_____

29. List the qualities unique to humans that have contributed to our success.

_____

_____

_____

_____

Skills Worksheet

Vocabulary Review

Complete each statement by writing the correct term or phrase in the space provided.

1. Something in the environment that causes an organism to react is a(n)

   _____, and the organism's reaction is its

   _____.

2. A(n) _____ is an action or series of actions performed by

   an animal in response to a stimulus.

3. Seasonal movement from breeding to feeding grounds is known as

   _____.

4. Innate behavior is called a(n) _____

   _____ _____ when the behavior

   always occurs in the same way.

5. The development of behaviors through experience is called

   _____.

6. A behavioral ritual that precedes mating is _____.

7. Behavior that increases reproductive success by helping to secure resources

   such as food and mates is known as _____

   _____.

8. Learning that can occur only during a specific period early in the life of an

   animal and cannot be changed once learned is called

   _____.

9. Behaviors for finding and gathering food are called

   _____ behaviors.

10. Learning by association is called _____.

11. Behaviors that occur on a daily basis are _____

_____.

12. Any behavior that contains information and involves a sender and a receiver is

_____.

13. The ability to draw a conclusion based on a fact or assumption is

_____.

14. Traits that increase the ability of individuals to attract or acquire mates are part

of an evolutionary mechanism called _____

_____.

15. Behavior that does not rely on experience in order to be produced is

_____ _____.

Skills Worksheet

Test Prep Pretest

In the space provided, write the letter of the term or phrase that best completes each statement or best answers each question.

_____ 1. Animals use signals to
    a. warn against predators.
    b. solicit play.
    c. attract a mate.
    d. All of the above

_____ 2. Scientists who question the reasons a behavior exists are asking
    a. a "how" question.
    b. a "why" question.
    c. about its evolution.
    d. Both (b) and (c)

_____ 3. Sexual selection is a(n)
    a. altruistic behavior.
    b. evolutionary mechanism.
    c. defensive behavior.
    d. territorial behavior.

_____ 4. In some animals, extreme traits for acquiring a mate include
    a. horns, antlers, and manes.
    b. the ability to learn.
    c. complex brain structure.
    d. All of the above

_____ 5. Which of the following is *not* a signal?
    a. feeding   c. color
    b. sound   d. scent

Question 6 refers to the figure at right.

_____ 6. The bird providing food to its young is engaging in
    a. foraging behavior.
    b. parental care.
    c. imprinting.
    d. territorial behavior.

_____ 7. Vocal communication is most developed in
      a. dogs.
      b. rodents.
      c. birds.
      d. primates.

In the space provided, write the letter of the description that best matches each term.

_____ 8. imprinting

_____ 9. foraging behavior

_____ 10. innate behavior

_____ 11. conditioning

_____ 12. territorial behavior

_____ 13. signal

_____ 14. operant conditioning

_____ 15. habituation

a. locating, obtaining, and consuming food

b. protecting a resource for exclusive use

c. occurs during a specific period early in an animal's life

d. used to influence another animal's behavior

e. ignoring a frequent, harmless stimulus

f. instinctive behavior

g. trial-and-error learning

h. learning by association

Complete each statement by writing the correct term or phrase in the space provided.

16. A(n) _____ is an action or series of actions performed by an animal in response to a stimulus.

17. To understand the factors that trigger or control a behavior, a scientist asks a(n) _____ question.

18. When new male lions in a pride kill cubs of other males, they are demonstrating a behavior influenced by _____

_____.

19. When rats locked in a box learned to depress a lever to get food, they demonstrated _____ _____ in a famous study conducted by B. F. Skinner.

20. Birds flying south for the winter are demonstrating _____

    _____.

21. When Pavlov's dogs learned to associate a ringing bell with meat powder

    which caused them to salivate, they demonstrated _____

    _____.

Read each question, and write your answer in the space provided.

22. Explain how imprinting in ducks and geese is influenced by both heredity
    and learning.

    _____

    _____

    _____

    _____

23. What is the difference between a "how" and a "why" behavioral question?
    Give an example of each.

    _____

    _____

    _____

    _____

24. List five types of signals and five methods animals can use to send and
    receive signals.

    _____

    _____

    _____

    _____

25. Explain the difference between habituation and classical conditioning.

    _____

    _____

    _____

    _____

Skills Worksheet

Vocabulary Review

In the space provided, write the letter of the description that best matches each term.

_____ 1. epithelial tissue

_____ 2. nervous tissue

_____ 3. muscle tissue

_____ 4. connective tissue

_____ 5. stem cells

a. carries information throughout the body

b. provides support, protection, and insulation

c. can develop into many different kinds of cells

d. cells contract and relax

e. lines most body surfaces

Complete each statement by writing the correct term or phrase in the space provided.

6. Red _____ _____ is soft tissue

inside bones that produces blood cells.

7. Bone cells called _____ maintain the mineral content of

bone.

8. A disease that results in large numbers of immature white blood cells is

_____, a cancer of tissues that produce blood cells.

9. A junction between two or more bones is called a(n) _____.

10. Bones of a joint are held together by strong bands of connective tissue called

_____.

In the space provided, explain how the terms in each pair differ in meaning.

11. ligament, tendon

_____

_____

12. leukemia, bone marrow

_____

_____

_____

Use the terms from the list below to fill in the blanks in the following passage.

actin	muscle fiber	sarcomere
extensor	myofibrils	tendons
flexor	myosin	

Most skeletal muscles are attached to bones by strips of dense connective tissue

called (13) _____. One muscle in a pair of muscles pulls a

bone in one direction, and the other muscle pulls the bone in the opposite direction.

A(n) (14) _____ muscle causes a joint to bend, and

a(n) (15) _____ muscle causes a joint to straighten.

Muscle tissue contains large amounts of protein filaments called

(16) _____ and (17) _____, which enable

muscles to contract. Each muscle cell, or (18) _____, is made

of small cylindrical structures called (19) _____, which have

alternating light and dark bands when viewed under a microscope. In the center of

each light band is a Z line, which anchors actin filaments. The area between the

two Z lines is called a(n) (20) _____, the functional unit of

muscle contraction.

In the space provided, write the letter of the description that best matches each term.

_____21. epidermis

_____22. keratin

_____23. melanin

_____24. dermis

_____25. subcutaneous tissue

_____26. sebum

a. protein that makes skin tough and waterproof

b. layer of skin that lies just beneath the epidermis

c. outer layer of skin

d. oily secretion that lubricates the skin

e. made mostly of fat

f. pigment that absorbs UV radiation

Skills Worksheet

Test Prep Pretest

In the space provided, write the letter of the term or phrase that best completes each statement or best answers each question.

_____ 1. From simplest to most complex, the four levels of structural organization of the human body are as follows:
  a. tissues, cells, organs, organ systems.
  b. cells, tissues, organs, organ systems.
  c. organ systems, tissues, cells, organs.
  d. cells, organs, tissues, organ systems.

_____ 2. Types of connective tissue include all of the following *except*
  a. blood.   c. fat.
  b. bone.   d. muscle.

_____ 3. Which of the following helps the body's temperature return to normal when you are cold?
  a. The body begins shivering to produce heat.
  b. The body increases blood flow to small vessels below the skin.
  c. The body begins secreting sweat on the surface of the skin.
  d. All of the above

_____ 4. The appendicular skeleton includes bones of the
  a. cranium.   c. pelvis.
  b. spine.   d. ribs.

_____ 5. In early development, bone tissue is made mostly of
  a. Haversian canals.   c. periosteum.
  b. bone marrow.   d. cartilage.

In the space provided, write the letter of the description that best matches each term.

_____ 6. compact bone

_____ 7. spongy bone

_____ 8. red bone marrow

_____ 9. yellow bone marrow

_____ 10. tendon

_____ 11. ligament

_____ 12. leukemia

_____ 13. osteocyte

a. maintains mineral content of bone

b. holds a movable joint together

c. site of blood cell production

d. cancer of the tissues that make blood cells

e. dense bone that provides a great deal of support

f. attaches a muscle to a bone

g. site of fat storage

h. loosely structured bone

Complete each statement by writing the correct term or phrase in the space provided.

14. The integumentary system includes the skin, _____, and

_____.

15. Osteoporosis can be delayed or prevented by a healthy

_____, regular _____, and

medication.

16. A muscle fiber contains many bundles of cylindrical structures called

_____.

17. When oxygen is plentiful, the ATP used to power muscle contractions is

supplied by _____ processes. As oxygen becomes

depleted, ATP is supplied by _____ processes.

18. A pigment found in the epidermis, _____, absorbs UV

radiation and causes tanning.

19. Hair consists mostly of dead cells filled with the protein called

_____, the same protein that makes the skin tough and

waterproof.

20. The dermis contains _____ _____,

which help regulate body temperature and bring nutrients to the skin's living

cells.

21. The most common type of skin cancer originates in cells of the

_____ that do not produce pigments.

22. Acne is caused by excessive secretion of _____, an oily

secretion that lubricates the skin, by oil glands.

Test Prep Pretest *continued*

Read each question, and write your answer in the space provided.

23. Describe the three main types of joints, and give an example of each.

_____

_____

_____

_____

_____

_____

_____

_____

24. Describe the interaction of myosin and actin during a muscle contraction.

_____

_____

_____

_____

_____

_____

25. Differentiate between the epidermis, the dermis, and subcutaneous tissue.

_____

_____

_____

_____

_____

_____

Name _____ Class _____ Date _____

Vocabulary Review

Use the terms from the list below to fill in the blanks in the following passage.

arteries   lymphatic system   red blood cells

capillaries   plasma   veins

cardiovascular system   platelets   white blood cells

The (1) _____ transports

materials throughout the body and distributes heat. Blood circulation describes the

route blood takes as it leaves and then returns to the heart. Blood vessels that carry

blood away from the heart are called (2) _____. From the

arteries, the blood passes into a network of smaller arteries called arterioles. From

arterioles, blood passes into (3) _____, which are tiny blood

vessels that allow the exchange of gases, nutrients, hormones, and other molecules

traveling in the blood. After leaving the capillaries, the blood flows into small

vessels called venules before emptying into larger vessels called (4)

_____, which are blood vessels that carry the blood back to

the heart.

About 55 percent of the total volume of blood is liquid called (5)

_____. Most of the cells that make up blood are

(6) _____

_____. The cells whose primary job is to defend the body

against disease are called (7) _____

_____. Cell fragments called (8)

_____ play an important role in the clotting of blood.

The (9) _____ is a system of

the body that collects and recycles fluids that leak from the capillaries. It is also

involved in fighting infections.

Vocabulary Review continued

In the space provided, write the letter of the description that best matches the term or phrase.

_____10. atrium

_____11. ventricle

_____12. platelets

_____13. plasma

_____14. red blood cells

_____15. white blood cells

_____16. blood pressure

_____17. pulse

_____18. heart attack

_____19. stroke

_____20. pharynx

_____21. larynx

_____22. trachea

_____23. bronchi

_____24. alveoli

_____25. diaphragm

a. blood cells that contain hemoglobin and carry oxygen

b. chamber that pumps blood away from the heart

c. blood cells that defend the body against disease

d. the force exerted by blood as it moves through the blood vessels

e. part of the blood that is mostly water with some nutrients, wastes, proteins, and salts mixed in

f. a rhythmic stretching of the blood vessels leading away from the heart

g. chamber that receives blood returning to the heart

h. condition that occurs when the blood supply to part of the heart is greatly reduced or stopped

i. cell pieces that pinch off from cells in bone marrow

j. condition that occurs when a blood vessel that carries oxygen and other materials to the brain bursts or is blocked by a blood clot

k. a long, straight tube in the chest cavity through which air passes

l. two small tubes that lead to the lungs

m. the voice box

n. a muscle below the rib cage that drives breathing

o. a muscular tube in the upper throat

p. air sacs where gases are exchanged

Skills Worksheet

Test Prep Pretest

In the space provided, write the letter of the term or phrase that best completes each statement or best answers each question.

_____ 1. The actual exchange of materials between the blood and the cells of the body occurs in the
   a. arteries.
   b. arterioles.
   c. veins.
   d. capillaries.

_____ 2. When fluids leak out of the cardiovascular system, they are returned by the
   a. respiratory system.
   b. lymphatic system.
   c. endocrine system.
   d. digestive system.

_____ 3. Blood type is determined by the presence or absence of
   a. A and B antigens dissolved in the blood plasma.
   b. A and O antigens on the surface of red blood cells.
   c. A and B antigens on the surface of red blood cells.
   d. A and B antigens on the surface of white blood cells.

_____ 4. Which of the following can cause arteries to harden and narrow?
   a. smoking
   b. high blood pressure
   c. high blood levels of fats
   d. All of the above

_____ 5. As the left ventricle contracts, the blood is prevented from moving back into the left atrium by
   a. a one-way valve.
   b. the superior vena cava.
   c. the inferior vena cava.
   d. the pulmonary veins.

_____ 6. Which of the following conditions is an inflammation of the lungs that can be caused by bacteria, viruses, or fungi?
   a. tuberculosis
   b. bronchitis
   c. pneumonia
   d. All of the above

| Test Prep Pretest *continued*

Complete each statement by writing the correct term or phrase in the space provided.

7. Carbon dioxide is an example of a(n) _____

   _____ transported by the cardiovascular system to the

   urinary system.

8. The heart receives blood in the two _____ and pumps

   blood away using the two _____.

9. The _____ _____ in the right atrium

   initiates each heart contraction.

10. The force that is exerted by blood as it moves through the blood vessels is

    called _____.

11. High blood pressure, or _____, can weaken the heart and

    damage blood vessels.

12. The _____ are suspended in the chest cavity and are

    bound on the sides by the ribs and on the bottom by the diaphragm.

13. The alveoli are connected to the bronchi by a network of tiny tubes called

    _____.

14. During breathing, _____ occurs when the diaphragm and

    rib cage return to their relaxed position.

15. Breathing rate is controlled by _____ in the brain and

    cardiovascular system.

16. Two diseases of the respiratory system that have been linked to cigarette

    smoking are _____ and _____

    _____.

17. When blood flow to the brain is stopped, tissues in the brain can be damaged

    or killed. This event is called a _____.

Read each question, and write your answer in the space provided.

18. List three risk factors for cardiovascular disease, and describe what you can do about them.

_____

_____

_____

_____

19. How does lymph move through the lymphatic system?

_____

_____

_____

_____

20. How are oxygen and carbon dioxide transported in the blood?

_____

_____

_____

_____

21. What is one factor that stimulates receptors in the brain, causing an increase in the breathing rate?

_____

_____

Name _____ Class _____ Date _____

Vocabulary Review

Complete each statement by writing the correct term or phrase in the space provided.

1. A(n) _____ is a substance the body needs for energy, growth, repair, and maintenance.

2. The process of breaking down food into molecules the body can use is called

   _____.

3. A(n) _____ is the amount of heat energy required to raise the temperature of 1 g of water 1°C (1.8°F).

4. Carbon-based substances that are necessary, in small amounts, for the normal metabolic functioning of the body are called _____.

5. Inorganic substances that are necessary to make certain body structures and substances, to continue normal nerve and muscle function, and to maintain osmotic balance are called _____.

In the space provided, write the letter of the description that best matches the term or phrase.

_____ 6. peristalsis

_____ 7. esophagus

_____ 8. pepsin

_____ 9. villi

_____ 10. urea

a. toxic metabolic waste

b. fine, fingerlike projections in the small intestine

c. a long tube that connects the mouth to the stomach

d. a digestive enzyme secreted by the stomach

e. wave of muscle contractions that pushes food into the stomach

| **Vocabulary Review** *continued*

Use the terms from the list below to fill in the blanks in the following passage.

excretion	ureters	urinary bladder
nephrons	urethra	urine

The process that rids the body of toxic metabolic wastes and that maintains

osmotic and pH balances is called (11) _____ . The organs

of excretion are the lungs, the kidneys, and the skin.

The tiny tubes in the kidneys with cup-shaped capsules surrounding a tight ball

of capillaries that filter wastes from the blood are (12) _____ . These

tubes retain useful molecules, and they produce (13) _____ . Urine

is carried from the kidneys into the (14) _____ _____by tubes called

(15) _____ . Urine leaves the body through a tube called the

(16) _____ .

Skills Worksheet

Test Prep Pretest

In the space provided, write the letter of the term or phrase that best completes each statement or best answers each question.

_____ 1. A substance needed by the body for energy, growth, repair, and maintenance is called a(n)
    a. fatty acid.   c. nutrient.
    b. simple sugar.   d. calorie.

_____ 2. All of the following are nutrients found in food *except*
    a. plasma.   c. proteins.
    b. carbohydrates.   d. vitamins.

_____ 3. A diet high in saturated fats can be linked to which of the following?
    a. kidney failure
    b. anorexia nervosa
    c. bulimia
    d. cardiovascular diseases

_____ 4. According to the MyPyramid food guidance system, a person should obtain most of their fat from
    a. beef, chicken, and fish.
    b. vegetable oils, nuts, and fish.
    c. fats, oils, and sweets.
    d. milk, yogurt, and cheese.

_____ 5. Amylases in saliva begin the breakdown of carbohydrates into
    a. fatty acids.   c. amino acids.
    b. polypeptides.   d. simple sugars.

_____ 6. In the stomach, single protein strands are cut into smaller amino acid chains by the digestive enzyme called
    a. amylase.   c. lipase.
    b. pepsin.   d. gastrin.

_____ 7. The products of digestion are absorbed into the bloodstream through the
    a. villi and microvilli of the small intestine.
    b. rectum of the large intestine.
    c. stomach and colon.
    d. liver and gallbladder.

_____ 8. Bile, which breaks fat globules into tiny fat droplets, is produced by the
    a. pancreas.   c. liver.
    b. gallbladder.   d. duodenum.

_____ 9. Which of the following is an example of chemical digestion?
  a. chewing food   c. breaking bonds
  b. peristaltic contractions   d. churning food

_____ 10. The end result of the filtration, reabsorption, and secretion processes in
  the nephrons is
  a. water.   c. urine.
  b. carbon dioxide.   d. urea.

_____ 11. Urine leaves the bladder and exits the body through a tube called the
  a. urethra.   c. kidney.
  b. ureter.   d. nephron.

Questions 12–14 refer to the figure at right.

_____ 12. The blood-filtering unit in the figure is
  called a(n)
  a. villus.
  b. nephron.
  c. urethra.
  d. microvillus.

_____ 13. The structure labeled A is called the
  a. collecting duct.
  b. glomerulus.
  c. renal tubule.
  d. Bowman's capsule.

_____ 14. The structure labeled C is called the
  a. collecting duct.
  b. glomerulus.
  c. renal tubule.
  d. Bowman's capsule.

Complete each statement by writing the correct term or phrase in the space provided.

15. The liver converts excess sugars to _____ and stores it
  for later.

16. Successive rhythmic waves of contraction of the smooth muscles around the

  esophagus, called _____ _____, move the food
  toward the stomach.

17. During digestion, the process of getting rid of undigested molecules and waste

    occurs in the _____ _____ .

18. The wall of the large intestine absorbs mostly _____

    _____ and _____ .

19. When you exhale, _____ _____ and

    some water are excreted by the lungs.

20. A procedure for filtering the blood called _____

    _____ can prolong the lives of many people with

    damaged kidneys.

Read each question, and write your answer in the space provided.

21. Describe the connection between heart disease and the MyPyramid food
    guidance system's recommendation for fats.

    _____

    _____

22. How do the liver and the pancreas differ from other digestive organs?

    _____

    _____

23. Describe the similarities and differences between a mineral and a vitamin.

    _____

    _____

24. Name two organs other than the kidney that are involved in excretion, and
    describe what each organ excretes.

    _____

    _____

25. Relate the role of water in maintaining a healthy body.

    _____

    _____

Skills Worksheet

Vocabulary Review

In the blanks provided, fill in the letters of the term or phrase being described.

1. a disease-causing agent

   _ **A** _ _ _ _ _ _

2. layers of epithelial tissue that serve as barriers to pathogens and produce chemical defense

   M _ _ _ _ _   **M** _ _ _ _ _ _ _

3. when chemicals and cells that attack pathogens gather around the site of injury or infection

   _ _ **L** _ _ _ _ _ _ _

4. chemical that causes local blood vessels to dilate

   _ _ _ **T** _ _ _ _ _

5. a white blood cell that activates the immune system

   _ _ _ **P** _ _ _ _ _ _ _ _

6. a white blood cell that destroys cells infected with pathogens

   _ _ _ _ _ _ _ _ **C** _ _ _ _ _ _

7. a white blood cell that makes proteins that bind to pathogens that are outside body cells

   _ **C** _ _ _

8. a white blood cell that ingests and kills pathogens

   _ _ _ **R** _ _ _ _ _

9. a white blood cell that makes and releases antibodies

   _ _ _ **S** _ _ . _ _ _ **L**

Use the terms from the list below to fill in the blanks in the following passage.

| antibodies | B cells | helper T cells |
| antigens | cytotoxic T cells | plasma cells |

   White blood cells are produced in bone marrow and circulate in blood and

lymph. Several kinds of white blood cells are involved in the immune response.

Macrophages consume pathogens and infected cells. The cells that attack and kill

infected cells are called (10) _____

_____. Cells called (11) _____ label

invaders for later destruction by macrophages. White blood cells that activate

Vocabulary Review *continued*

both cytotoxic T cells and B cells are (12) _____

_____.

An infected body cell will display (13) _____ of an

invader on its surface. These are substances that trigger an immune response.

Activated B cells produce (14) _____

_____, which release Y-shaped proteins into the blood.

These special proteins are called (15) _____.

In the space provided, write the letter of the description that best matches each term.

_____16. allergen

_____17. immunity

_____18. memory cell

_____19. vaccine

_____20. antibody

_____21. autoimmune disease

_____22. AIDS

_____23. HIV

_____24. allergy

a. body's response to a normally harmless antigen

b. when the body launches an immune response against its own cells

c. a white blood cell that protects the body from pathogens the body has already encountered

d. the virus that causes AIDS

e   weak antigens that most people do not react to

f. long-lasting resistance to a particular disease

g. a solution that contains a dead or modified pathogen that can no longer cause disease

h. acquired immunodeficiency syndrome

i. protein that binds to a specific antigen

Skills Worksheet

Test Prep Pretest

In the space provided, write the letter of the term or phrase that best completes each statement or best answers each question.

_____ 1. White blood cells that kill bacteria by engulfing them and then
releasing chemicals that kill both the bacteria and themselves are
   a. macrophages.   c. pathogens.
   b. memory cells.   d. helper T cells.

_____ 2. Fever is helpful in fighting bacteria because
   a. higher temperatures promote the activation of cellular proteins.
   b. lower temperatures promote the activation of cellular proteins.
   c. disease-causing bacteria do not grow well at high temperatures.
   d. disease-causing bacteria do not grow well at low temperatures.

_____ 3. Which kind of white blood cell activates a specific immune response?
   a. cytotoxic T cells   c. helper T cells
   b. natural killer cells   d. All of the above

Questions 4–7 refer to the figure at right, which shows an immune response.

_____ 4. The cell labeled *A* is a
   a. macrophage.
   b. helper T cell.
   c. cytotoxic T cell.
   d. B cell.

_____ 5. The cell labeled *B* is a
   a. plasma cell.
   b. cytotoxic T cell.
   c. helper T cell.
   d. B cell.

_____ 6. The cell labeled *C* is a
   a. cytotoxic T cell.
   b. helper T cell.
   c. plasma cell.
   d. memory cell.

_____ 7. The cells produced by
the cell labeled *D*
   a. release antibodies.
   b. kill infected cells.
   c. engulf viruses.
   d. infect body cells.

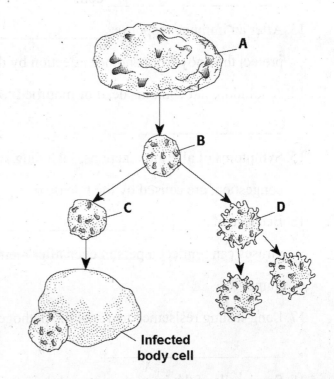

Infected
body cell

| Test Prep Pretest *continued*

_____ 8. In an autoimmune disease,
        a. a pathogen is immune to antigens.
        b. a pathogen circulates in the blood.
        c. the body attacks its own cells.
        d. All of the above

_____ 9. Which of the following is *not* a way that HIV can be transmitted?
        a. semen
        b. blood
        c. breast milk
        d. insect bites

Complete each statement by writing the correct term or phrase in the space provided.

10. Membranes lining the respiratory tract secrete a layer of

_____ that traps pathogens before they can enter the lungs.

11. Cells infected with viruses release proteins called _____,

which stop viruses from making proteins and RNA.

12. Helper T cells activate _____ T cells and

_____ cells.

13. After an immune response, _____ cells circulate in the blood and

protect the body from another infection by the same pathogen.

14. A solution that contains dead or modified forms of a pathogen is called a(n)

_____.

15. Symptoms of allergic reactions, including swelling, itchy eyes, and nasal

congestion, are caused by the release of _____.

16. Because of _____ _____, influenza

viruses can reinfect a person even after memory cells have produced

immunity.

17. Long-lasting resistance to a specific pathogen is called

_____.

18. Some cells of the immune system have receptor proteins that bind to specific

_____.

Test Prep Pretest *continued*

Read each question, and write your answer in the space provided.

19. List and describe two ways the body prevents pathogens from entering it.

_____

_____

_____

20. What is the difference between an antigen and an antibody?

_____

_____

_____

21. What causes the pus that accompanies some infections?

_____

_____

_____

22. Name three kinds of white blood cells involved in inflammation. How does each type attack pathogens?

_____

_____

_____

_____

_____

23. How does a person become immune to a pathogen?

_____

_____

_____

_____

_____

_____

_____

24. Name three autoimmune diseases and describe their symptoms.

_____

_____

_____

_____

_____

_____

_____

25. Describe the connection between HIV infection and a weakened immune
system.

_____

_____

_____

_____

_____

Name _____ Class _____ Date _____

Skills Worksheet

Vocabulary Review

In the space provided, write the letter of the description that best matches each term.

_____ 1. neuron

_____ 2. dendrite

_____ 3. axon

_____ 4. nerve

_____ 5. membrane potential

_____ 6. action potential

_____ 7. synapse

_____ 8. neurotransmitter

a. the difference in electrical charge across a cell membrane

b. conducts nerve impulses away from the cell body

c. nerve impulse

d. nerve cell; transmits information throughout the body

e. bundle of axons

f. part of a neuron that receives information from other neurons

g. a junction at which a neuron meets another cell

h. a chemical that transmits nerve impulses across synapses

Write the correct term from the list below in the space next to its definition.

brain	cerebellum	reflex
brainstem	cerebrum	spinal cord
central nervous system	peripheral nervous system	

_____ 9. the largest part of the brain, which controls most sensory and motor processing

_____ 10. consists of the brain and spinal cord

_____ 11. dense cable of nervous tissue that runs through the vertebral column

_____ 12. contains sensory neurons and motor neurons

_____ 13. the body's main processing center

_____ 14. collection of structures leading down to the spinal cord

_____ 15. regulates balance, posture, and movement

_____ 16. a sudden, involuntary contraction of muscles in response to a stimulus

| Vocabulary Review *continued*

In the space provided, write the letter of the description that best matches each term.

_____17. sensory receptor

_____18. retina

_____19. taste buds

_____20. cochlea

_____21. semicircular canal

a. the lining of photoreceptors and neurons in the eye
b. aids in hearing
c. a specialized neuron that detects sensory stimuli
d. helps maintain equilibrium
e. detect sugars, acids, alkaloids, and salts, and proteins

Complete each statement by writing the correct term or phrase in the space provided.

22. The need for increasing amounts of a drug to achieve the desired sensation is

    called _____.

23. A drug that generally decreases the activity of the central nervous system is

    called a(n) _____.

24. A drug that generally increases the activity of the central nervous system is

    called a(n) _____.

25. Drugs that alter the functioning of the central nervous system are known as

    _____ _____.

26. A set of emotional and physical symptoms caused by removing a drug from

    the body of a drug addict is _____.

27. A physiological response caused by use of a drug that alters the normal

    functioning of neurons and synapses is _____.

Skills Worksheet

Test Prep Pretest

In the space provided, write the letter of the term or phrase that best completes each statement or best answers each question.

_____ 1. During a knee-jerk reflex, the nerve impulse is received by the
   a. brain.
   b. spinal cord.
   c. spinal cord and then the brain.
   d. thalamus.

Questions 2–4 refer to the figure below, which shows the structure of a typical neuron.

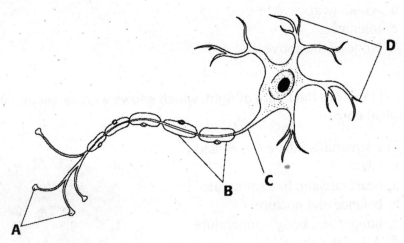

_____ 2. What occurs when an action potential reaches the structures labeled *A*?
   a. Neurotransmitters are released.
   b. Myelin sheaths are activated.
   c. Cell bodies receive messages.
   d. All of the above

_____ 3. The structures labeled *B* are
   a. axon terminals.   c. dendrites.
   b. nodes of Ranvier.   d. myelin sheaths.

_____ 4. The structures labeled *D* are
   a. dendrites.   c. axon terminals.
   b. axons.   d. nodes of Ranvier.

_____ 5. Light entering the eye stimulates
   a. hair cells in the retina.
   b. the optic nerve.
   c. rods and cones in the retina.
   d. mechanoreceptors.

_____ 6. When cocaine interferes with reuptake receptors on a presynaptic neuron, the
   a. postsynaptic cell is overstimulated.
   b. number of neurotransmitter receptors decreases.
   c. excess neurotransmitters remain in the synaptic cleft.
   d. All of the above

_____ 7. Drug use that alters normal functioning of neurons and synapses results in
   a. addiction.
   b. withdrawal.
   c. tolerance.
   d. None of the above

Questions 8–11 refer to the figure at right, which shows a cross section of the brain and spinal cord.

_____ 8. The structures labeled *A, B,* and *C* regulate
   a. heart rate and breathing rate.
   b. balance and posture.
   c. hunger and body temperature.
   d. All of the above

_____ 9. The structure labeled *D* is involved in
   a. balance and posture.
   b. maintaining homeostasis.
   c. sensory and motor processing.
   d. spinal reflexes.

_____ 10. The structure labeled *E* is the
   a. thalamus.
   b. corpus callosum.
   c. brain stem.
   d. cerebrum.

_____ 11. The structure labeled *F* is the
   a. thalamus.
   b. hypothalamus.
   c. cerebellum.
   d. cerebral cortex.

| Test Prep Pretest *continued*

Complete each statement by writing the correct term or phrase in the space provided.

12. The _____ _____ of a neuron is

negative because there are more positively charged ions outside the cell than

inside the cell.

13. A(n) _____ _____ is a sudden

change in the polarity of a neuron's cell membrane.

14. During synaptic transmission, a presynaptic neuron releases a(n)

_____ into the synaptic _____.

15. At a synapse, a neurotransmitter may _____ or

_____ the activity of the postsynaptic cell to which

it binds.

16. After a nerve impulse has passed, _____ ions flow out of

the axon, and the membrane potential becomes _____

again.

17. In the spinal cord, cell bodies of neurons make up _____

matter, whereas axons make up _____ matter.

18. The _____ nervous system contains neurons that connect

the brain and the spinal cord to the rest of the body.

19. The coiled inner ear structure that converts sound waves to nerve impulses is

called the _____.

20. Auditory information is processed in the _____

_____ of the brain.

21. Drugs that alter the functioning of the central nervous system and are often

addictive are called _____ _____.

22. A(n) _____ is a substance that decreases the activity of

the central nervous system.

Test Prep Pretest *continued*

Read each question, and write your answer in the space provided.

23. Distinguish between the somatic nervous system and the autonomic nervous system.

_____

_____

_____

_____

_____

24. List the five types of chemicals that taste buds can detect.

_____

_____

_____

25. List three ways the nervous system can be damaged.

_____

_____

_____

26. Why is it important to wear a seat belt in the car and a helmet when riding a bicycle, skateboarding, or in-line skating?

_____

_____

_____

_____

_____

Skills Worksheet

Vocabulary Review

In the space provided, write the letter of the description that best matches each term.

_____ 1. hormones

_____ 2. endocrine glands

_____ 3. target cell

_____ 4. antagonistic hormone

_____ 5. second messenger

_____ 6. feedback mechanism

_____ 7. androgens

_____ 8. progesterone

a. female sex hormone

b. a molecule that passes a message from a hormone to the inside of a cell

c. a hormone that counteracts the effect of another hormone

d. a specific cell on which a hormone acts

e. male sex hormones

f. substances that are made in one part of the body and cause changes in another part of the body

g. a system in which one step in a series of events controls an earlier step

h. ductless glands that secrete hormones directly into either the bloodstream or the fluid around cells

In the space provided, explain how the terms in each pair are related to each other.

9. epinephrine, norepinephrine

_____

_____

_____

10. estrogen, androgens

_____

_____

_____

_____

Skills Worksheet

Test Prep Pretest

In the space provided, write the letter of the term or phrase that best completes each statement or best answers each question.

_____ 1. Which of the following is *not* a characteristic of the endocrine system?
a. Its chemical messengers are neurotransmitters.
b. It coordinates all of the body's sources of hormones.
c. Endocrine cells can release hormones directly into the bloodstream.
d. Its chemical messengers bind to receptors.

_____ 2. Which of the following is a function of hormones?
a. regulating growth
b. maintaining homeostasis
c. reacting to stimuli
d. All of the above

_____ 3. A hormone acts only on its target cell by
a. stimulating a nerve cell.
b. binding to a target cell's receptor protein.
c. binding to a nerve cell.
d. activating an enzyme in the blood.

_____ 4. The testes produce
a. estrogen.
b. progesterone.
c. androgens.
d. All of the above

_____ 5. Epinephrine and norepinephrine are released in response to
a. low blood-calcium levels.
b. stress.
c. darkness.
d. high blood-glucose levels.

_____ 6. Aldosterone helps to
a. excrete sodium ions in the urine.
b. regulate salt concentrations in the blood.
c. decrease blood pressure.
d. regulate calcium concentrations in the blood.

_____ 7. Which of the following hormones is secreted by the parathyroid gland?
a. parathyroid hormone
b. prolactin
c. thyroid hormone
d. All of the above

Test Prep Pretest *continued*

Complete each statement by writing the correct term or phrase in the space provided.

8. Ductless organs that produce hormones are called _____

_____.

9. Hormones that are fat-soluble are _____ hormones.

10. The release of _____ from the hypothalamus stimulates

uterine contractions during childbirth.

11. When a steroid hormone binds to a receptor protein in a target cell's

cytoplasm, a(n) _____ - _____

complex is produced.

12. In a(n) _____ _____ mechanism,

high levels of a hormone inhibit the output of more hormone.

13. A high level of the adrenal cortex hormone, called _____,

suppresses the immune system.

14. The _____ and the _____ gland

together serve as a major control center for the rest of the endocrine system.

15. A child who has _____ may have stunted growth and

brain damage.

16. Hormones that regulate the body's metabolic rate and are involved in growth,

development, and reproduction are _____ hormones.

17. The _____ _____ secretes the

hormone melatonin in response to darkness.

18. The hormone _____ regulates the body's daily sleep

cycle.

19. Hormonelike substances called _____ are made where

tissues are injured and cause pain and inflammation.

Read each question, and write your answer in the space provided.

20. Describe the role of second messengers in relaying a hormone's message.

_____

_____

_____

_____

_____

_____

21. Explain how the pancreatic hormones insulin and glucagon regulate blood-glucose levels.

_____

_____

_____

_____

_____

_____

22. Describe three basic ways that the endocrine system can malfunction.

_____

_____

_____

_____

_____

_____

Name _____ Class _____ Date _____

Vocabulary Review

Write the correct term from the list below in the space next to its definition.

epididymis	semen	tubule
penis	seminiferous	vas deferens
prostate gland	testis	

_____ 1. long tube that connects the epididymis to the urethra

_____ 2. one of the gamete-producing organs of the male reproductive system

_____ 3. secretes an alkaline fluid that neutralizes the acids in the female reproductive system

_____ 4. a mixture of secreted fluids and sperm

_____ 5. one of the many tightly coiled tubes within the testes where sperm are produced

_____ 6. the male organ that deposits sperm in the female reproductive system during sexual intercourse

_____ 7. long, coiled tube where sperm mature

In the space provided, write the letter of the description that best matches each term.

_____ 8. ovulation

_____ 9. embryo

_____ 10. fetus

_____ 11. ovum

_____ 12. implantation

_____ 13. menstruation

a. a developing human during the first 8 weeks after first cleavage

b. the release of a mature egg cell from the organ in which egg cells are produced

c. the shedding of the lining of lining of the uterus that occurs on a cyclical basis

d. a developing human from the eighth week of pregnancy until birth

e. the burrowing of the blastocyst into the lining of the uterus

f. a mature egg cell

| Vocabulary Review *continued*

Complete *each* statement by writing the correct term or phrase in the space provided.

14. A(n) _____ is one of the gamete-producing organs of the female reproductive system.

15. The _____ is the hollow, muscular organ in which an embryo embeds itself and grows and develops.

16. The _____ is the muscular tube that leads from the outside of a female's body to the uterus.

17. Each _____ _____ is a passageway through which an ovum moves from an ovary toward the uterus.

18. The series of changes that prepare the uterus for a possible pregnancy each month is called the _____ _____.

19. A common causes of infertility in women is _____ _____ _____, which is a severe inflammation of the upper reproductive system, including the uterus, ovaries, and fallopian tubes, that results from an untreated bacterial STI.

20. A very common viral STI that includes periodic outbreaks of painful blisters in the genital region and flulike aches and fever is _____ _____.

Name _____ Class _____ Date _____

Skills Worksheet

Test Prep Pretest

In the space provided, write the letter of the term or phrase that best completes each statement or best answers each question.

_____ 1. Sperm cells are produced by meiosis in the
    a. epididymis.
    b. vas deferens.
    c. seminiferous tubules.
    d. prostate gland.

_____ 2. The gamete-producing organs of the female reproductive system are the
    a. corpus luteum.   c. ovaries.
    b. fallopian tubes.   d. seminiferous tubules.

_____ 3. When the zygote reaches the uterus, it is a hollow ball of cells called a(n)
    a. ovum.   c. follicle.
    b. blastocyst.   d. fetus.

_____ 4. Which process occurs in the female body when an egg is *not* fertilized?
    a. ejaculation   c. ovulation
    b. implantation   d. menstruation

_____ 5. Which event occurs at the end of the third trimester of pregnancy?
    a. The uterine walls contract and expel the fetus.
    b. The chorion and uterus interact to form the placenta.
    c. A fetus becomes an embryo.
    d. A sperm cell penetrates an ovum.

_____ 6. Sexually transmitted infections (STIs) can be caused by
    a. bacteria.   c. viruses.
    b. protists.   d. All of the above

Complete each statement by writing the correct term in the space provided.

7. A fatal disease caused by the human immunodeficiency virus (HIV) is

_____.

8. The only sure way to protect yourself against contracting a sexually-

transmitted infection is to practice _____.

| Test Prep Pretest *continued*

Questions 9 and 10 refer to the figure below, which shows a mature sperm cell.

Complete each statement by writing the correct term in the space provided.

9. The structure labeled *A,* called the _____ of the sperm, contains enzymes that help the sperm penetrate an ovum.

10. The energy that sperm need for movement is supplied by mitochondria in the _____, labeled *B*. This energy powers the whiplike movements of the _____, labeled *C*.

11. The testes are located outside the body cavity in an external skin sac called the _____.

12. When a follicle bursts, the mature egg cell is released in a process called _____.

13. During the luteal phase of the ovarian cycle, the lining of the _____ thickens and becomes filled with fluids and nutrients.

14. During _____, the uterine lining is shed, blood vessels are broken, and a mixture of blood and tissues leave the body through the _____.

15. _____ and _____ occur before implantation of the blastocyst.

16. Before the end of it's eighth week, a developing human is a(n) _____; afterward, it is a(n) _____.

17. By the end of the _____ trimester, a fetus is able to survive outside the mother's body.

18. Two bacterial sexually transmitted infections (STIs) that often cause pelvic inflammatory disease (PID) are _____ and _____.

| Test Prep Pretest *continued* |

Read each question, and write your answer in the space provided.

19. Trace the path that sperm travel once they leave the testes.

_____

_____

_____

_____

_____

_____

20. Describe the events that occur early in the first trimester of pregnancy.

_____

_____

_____

_____

Question 21 refers to the figure at right, which shows the female reproductive system.

21. Identify the structures labeled *A–D*, and describe the functions of these structures.

_____

_____

_____

_____

_____

_____

_____

_____

_____

_____

_____

Skills Worksheet

Vocabulary Review

In the space provided, write the letter of the description that best matches each term.

_____ 1. pathology

_____ 2. chromatograph

_____ 3. forensic science

_____ 4. ballistics

_____ 5. autopsy

a. tool that separates chemicals based on their physical properties

b. scientific study of disease

c. examination of a dead body

d. science that deals with the motion and impact of projectiles

e. the use of science to investigate legal matters

Complete each statement by writing the correct term or phrase in the space provided.

6. The stiffening of muscles after a person dies is called _____

_____.

7. A(n) _____ is a tool that records how a substance reacts with

wavelengths of electromagnetic radiation.

8. _____ is the settling of blood to

the lowest points of the body after death.

9. The study of harmful substances, called _____, and their

effects on the body is _____.

10. _____ _____, the cooling of the body

after death, occurs for about 24 hours.

Skills Worksheet

Test Prep Pretest

In the space provided, write the letter of the term or phrase that best completes each statement or best answers each question.

_____ 1. The Locard exchange principle states that
  a. when a gun is fired, residue is produced.
  b. forensic scientists exchange testimony for evidence.
  c. tissues may not be removed from a body during an autopsy.
  d. evidence is exchanged whenever two people come in contact with each other.

_____ 2. Which of the following is *not* a duty of a forensic scientist?
  a. arrest suspects
  b. testify in court
  c. analyze evidence
  d. determine the cause of accidents

_____ 3. Which tool would a forensic scientist use to find out if a gun has been used in more than one crime?
  a. computer   c. spectrometer
  b. microscope   d. chromatograph

_____ 4. Which of the following is a meaning of the word *identity* in forensics?
  a. What is it?   c. Where is it?
  b. Who did it?   d. None of the above

_____ 5. A chromatograph separates substances based on their
  a. appearance.   c. chemical properties.
  b. physical properties. d. interactions with light.

In the space provided, write the letter of the description that best matches each term.

_____ 6. ballistics

_____ 7. autopsy

_____ 8. pathology

_____ 9. toxicology

_____ 10. PCR

_____ 11. fingerprint

a. used to make many copies of DNA

b. one type of friction ridges

c. the scientific study of harmful substances and their effects on the body

d. deals with the motion and impact of projectiles

e. the scientific study of disease

f. examination of a body after death

Complete each statement by writing the correct term or phrase in the space provided.

12. Forensic toxicology has two branches: _____ toxicologists

    analyze samples from living persons, whereas _____

    toxicologists analyze samples from dead persons.

13. Two unique characteristics that are used to identify a person are

    _____ and _____

    _____.

14. Hair, fibers, glass, paint, and pollen are examples of _____

    _____.

15. Marks made by a crowbar to break into a car are called

    _____.

16. Rigor mortis is the stiffening of _____ that occurs after

    death.

17. _____ _____ is the cooling of the body

    that occurs after death.

18. _____ _____ is the settling of blood to

    the lowest points of the body after death.

19. _____ _____ cause a body to warm up about 24 hours

    after death.

20. The identities of the _____ and the

    _____ must be discovered in order to solve a crime.

21. Forensic entomologists apply the study of _____ to legal

    matters.

Test Prep Pretest *continued*

22. What are the steps of an investigation at a crime scene?

_____

_____

_____

23. What are the three components that describe how a person died? Give an example
of each component.

_____

_____

_____

_____

24. What information is used to estimate the time of death?

_____

_____

_____

_____

25. Describe two uses of DNA analysis in addition to solving crimes.

_____

_____

_____